Inversion and Data Assimilation in Remote Sensing

SCIENCES

Image, Field Director – Laure Blanc-Féraud

Remote Sensing Imagery,
Subject Heads – Emmanuel Trouvé and Avik Bhattacharya

Inversion and Data Assimilation in Remote Sensing

Estimation of Geophysical Parameters

Coordinated by
Yajing Yan

iSTE

WILEY

First published 2024 in Great Britain and the United States by ISTE Ltd and John Wiley & Sons, Inc.

ISTE Ltd
27-37 St George's Road
London SW19 4EU
UK

www.iste.co.uk

John Wiley & Sons, Inc.
111 River Street
Hoboken, NJ 07030
USA

www.wiley.com

Library of Congress Control Number: 2024933649

British Library Cataloguing-in-Publication Data
A CIP record for this book is available from the British Library
ISBN 978-1-78945-142-9

ERC code:
PE10 Earth System Science
PE10_14 Earth observations from space/remote sensing

Contents

Preface

Yajing YAN
LISTIC, Université Savoie Mont-Blanc, Annecy, France

This book is part of the SCIENCES series by ISTE–Wiley and belongs to the Image field of the Engineering and Systems department. The field, Image, covers the entire processing chain, from acquisition to interpretation, by analyzing the data provided by various imaging systems. This field is split into seven subjects, including remote sensing imagery (RSI). Based on this subject, we propose a series of books that portray diverse and comprehensive topics in advanced remote sensing images and their applications for earth observation (EO). There has been an increasing demand for monitoring and predicting our planet's evolution on a local, regional and global scale. Hence, airborne, space-borne and ground-based platforms with active and passive sensors have acquired images that measure several features at various spatial and temporal resolutions over the past few decades.

RSI has become a broad multidisciplinary domain, attracting scientists across the diverse fields of science and engineering. The books proposed in this RSI series aim to present state-of-the-art and available scientific knowledge about the primary sources of images acquired by optical and radar sensors. The books will cover the processing methods that are developed by the signal and image processing community to extract useful information for end-users for an extensive range of EO applications. Each RSI book focuses on general topics such as change detection, surface displacement measurement, target detection, model inversion and data assimilation.

This book addresses recent advances in remote sensing data assimilation and inversion. Besides a summary of the classical methods, it deals with a wide range of themes where assimilation and inversion techniques have been developed and/or adapted according to the domain specificity; that is, in terms of scientific objectives, characteristics of data and physical models. This book is organized into two parts:

assimilation and inversion. Each part begins with a methodological chapter that presents and summarizes the most widely used methods and follows with chapters of various thematic applications. The assimilation part focuses on prediction problems by assimilation of in situ and remote sensing data in land surface monitoring, wildfire modeling, volcanology and glaciology, with the methodological barrier related to the specificity of each theme being highlighted and discussions on possible solutions to explore in order to better implement data assimilation techniques being opened. The inversion part includes biomass inversion with SAR images, bio-physical-hydrological inversion in coast zones with multi- and hyper-spectral images. Moreover, very recent advent in solving inverse problems by neural networks is showcased with illustrations in astrophysical applications.

This book is intended for students, engineers and researchers who want to gain broad knowledge of inversion and data assimilation in the domain of remote sensing imagery. This book has been written with a readership that is acquainted with applied mathematics, signal processing, remote sensing and geo-science. It should enable students, engineers and researchers in these communities to benefit from the methods and results presented in the book. This book is not dedicated to specific themes presented, but the references given will allow readers who want to delve into the subject to refer to other related works. For further in-depth mathematical tools in imaging inverse and optimization problems, readers can refer to "Mathematical tools for solving imaging inverse problems" (edited by P. Escande, D. Fortun and E. Soubies) and "Optimization for Imaging Sciences" (edited by E. Chouzenoux, J.C. Pesquet, N. Pustelnik and A. Repetti), two incoming books in the Image field.

This book represents some collective effort and contributions from a large number of authors. I hereby thank all contributors again. We would also like to thank Emmanuel Trouvé and Avik Bhattacharya who initiated and coordinated the book series "Remote Sensing Imagery" to which the current book belongs.

Yajing YAN
June 2024

PART 1

Data Assimilation

1

Methods for Assimilation of Observations: Application to Numerical Weather Prediction

Olivier TALAGRAND

LMD/IPSL, CNRS, ENS, Université PSL, Institut polytechnique de Paris, Sorbonne Université, Paris, France

1.1. Introduction

This chapter presents a general description of *assimilation of observations*. The guideline of the presentation is the development of assimilation in the context of numerical weather prediction. The case of numerical weather prediction is rather emblematic in that it has led to the gradual development of more and more powerful numerical algorithms for assimilation of observations, which have in turn significantly contributed to the improvement of weather prediction, while they were at the same time expanding to many diverse applications.

Weather prediction, like many other fields of present science and technology, is now largely dominated by numerical modeling. Meteorological forecasts are basically produced by numerical models which, started from the observed meteorological conditions, compute the expected evolution of the relevant parameters (pressure, temperature, wind and humidity) over periods which, depending on the particular purpose at hand, may vary from a few hours or less, to weeks or months. The numerical models are built on the physical laws that govern the evolution of the

Inversion and Data Assimilation in Remote Sensing,
coordinated by Yajing YAN. © ISTE Ltd 2024.

flow (namely, the laws of conservation of mass, energy and momentum, or at least as accurate a formulation of those laws as can be obtained). These laws are discretized on an appropriate three-dimensional spatial grid and integrated in time over the period of the forecast. As an example, the present operational model of the European Centre for Medium-Range Weather Forecasts (ECMWF), located in Reading in England, is built on a grid which covers the whole volume of the atmosphere, with horizontal and vertical resolutions of about 9 km and 2 km, respectively. This leads for the description of the instantaneous state of the whole atmosphere to a vector with dimension over 10^9. Two daily forecasts are produced at present by evolving the state vector over a 10-day period (with a time step of 450 s).

With regard to the initial conditions from which the forecasts must be started, the observing system consists of a large variety of different instruments. Radiosondes are launched from ground stations at conventional synchronous "synoptic" hours. The ground stations are, however, concentrated over continents, mostly in developed countries. Satellites measure the radiation emitted by the planet, providing information on the thermal structure of the atmosphere. They cover the surface of the Earth more homogeneously than ground stations, but perform observations continuously in time. Other types of observations (as performed, for instance, by commercial planes or drifting buoys at the surface of the ocean) are also available. At present, the number of scalar observed quantities that are used in operational weather prediction over a 24-h period lies in the range 10^7–10^8.

A large fraction of these observations is distributed over time, and there is much more information in the observations distributed over 12 or 24 h, over which the flow can significantly evolve, than in synoptic observations performed at a given time. *Assimilation* is the process through which all that information is combined in order to define the initial conditions of the forecast. Its purpose can be precisely defined as estimating the state of the observed system as accurately as possible using all available relevant information. That information essentially consists of two different parts. The observations stricto sensu, on the one hand, may vary in nature, accuracy, as well as in spatial and temporal resolution and distribution, and the physical laws that govern the flow, on the other hand, are available in practice in the form of a numerical and necessarily approximate model.

Assimilation is one of many inverse problems that are encountered in many fields of science and technology. Inverse problems arise in situations when we want to know the state of a system (often a physical system, but not necessarily so), while the available data are not in the format appropriate for the purpose at hand. They are encountered, for instance, in solid Earth geophysics (when we want to know the internal structure of the planet), non-destructive probing of structures of various types, navigation of aircraft, spacecraft or any mobile object, as well as in many other applications as described in the other chapters of the book. In most of these problems, data are affected with some uncertainty, and we may wish to know the resulting uncertainty on the final estimate. From a mathematical point of view, uncertainty is conveniently described by probability distributions. This leads to the fact that many inverse problems are stated as problems in Bayesian estimation, namely determining the probability distribution for the state of the system of interest, conditioned by the available data. Indeed, although the basic nature of the problems considered may be very different, the mathematical equations that are used for solving these problems are often the same (as can be seen in several chapters of this book).

It is in this general Bayesian perspective that assimilation of meteorological observations is described and discussed below. Compared to other inverse problems, there are two specific difficulties in assimilation of meteorological observations. Firstly, the observations used are distributed in time, so that the complex nonlinear dynamics of the atmospheric flow, and in particular the instabilities that constantly develop in the flow, must be taken into account. Secondly, as shown by the numbers given above, the numerical size of the problems to be solved is extremely large. This difficulty is aggravated by the need for the forecast that will follow to be delivered in time. Those two specific difficulties have had a very strong impact on the development of assimilation of meteorological observations.

Oceanography is another domain where assimilation of observations has been used early. The dynamics of the ocean is very similar to that of the atmosphere, with different spatial and temporal scales. Due to its opacity to any form of electromagnetic radiation, the ocean is much more difficult to observe than the atmosphere. However, similar problems arise as in meteorology, and much of what is said below also applies to the ocean. Actually, assimilation of oceanographic observations is at the origin of several of the developments that are described in the following.

Section 1.2 presents a classical and relatively simple problem in Bayesian estimation in the linear and Gaussian case. The relatively simple algorithm called *optimal interpolation* (OI) is described in section 1.3 as an application of that elementary linear and Gaussian approach. The main algorithms that are used in operational assimilation, namely, *Variational Assimilation* and *Ensemble Kalman Filter* (EnKF), are presented and discussed as extensions of the same approach in section 1.4. A new class of algorithms, namely *Particle Filters*, is then described and discussed (section 1.5). The new methods of artificial intelligence (AI) and machine learning (ML) are briefly presented in section 1.6, and various extensions and applications of assimilation are described in section 1.7.

Early studies on assimilation of meteorological and oceanographical observations are as follows: Daley (1991); Ghil and Malanotte-Rizzoli (1991); Talagrand (1997);

Kalnay (2002). More recent studies are as follows: Asch et al. (2016) and Carrassi et al. (2018).

In the following, $E()$ denotes the expectation of a (scalar or vector) random variable. $\mathcal{N}(a, \mathbf{C})$ denotes the multidimensional Gaussian probability distribution with expectation a and covariance matrix $\mathbf{C}$ (with a similar transparent notation for a uni-dimensional Gaussian probability distribution).

1.2. The linear and Gaussian case

A simple estimation problem is as follows: we want to determine the unknown *state vector* x of the system of interest, belonging to *state space* $\mathcal{S}$, with dimension n. The known available data are supposed to make up a vector z, belonging to *data space* $\mathcal{D}$, with dimension m. The data vector is related to the unknown x through the relationship:

$$z = \mathbf{\Gamma}x + \zeta \qquad [1.1]$$

where the *data operator* $\mathbf{\Gamma}$ is a known linear operator from state space into data space, and ζ is an unknown "error", meant to represent the effect of all possible uncertainties in the link between x and z (instrumental and/or representativeness errors, inaccuracies in the specification of $\mathbf{\Gamma}$, etc.). Assume that the error ζ is a random Gaussian variable, with expectation 0 and covariance matrix $\mathbf{S}$ ($\zeta \sim \mathcal{N}(0, \mathbf{S})$) (if the expectation of ζ was not 0, but known, it would have to be first subtracted from z). The probability distribution of x, conditioned to z, is then the Gaussian distribution $\mathcal{N}(x^a, \mathbf{P}^a)$, with:

$$x^a = (\mathbf{\Gamma}^T\mathbf{S}^{-1}\mathbf{\Gamma})^{-1}\mathbf{\Gamma}^T\mathbf{S}^{-1}z \qquad [1.2]$$

$$\mathbf{P}^a = E[(x^a - x)(x^a - x)^T] = (\mathbf{\Gamma}^T\mathbf{S}^{-1}\mathbf{\Gamma})^{-1} \qquad [1.3]$$

These quantities are properly defined under the only condition that rank$\mathbf{\Gamma} = m$. This is a *determinacy* condition, which means that z contains information, either directly or indirectly, on every component of x. It requires $m \geq n$. We will set $p = m - n$. The upper index a in equations [1.2] and [1.3] refers to the fact that, in the standard vocabulary of meteorology, the estimate x^a is called the *analysis* of the state of the system. The expectation $E[(x^a - x)]$ of the error in x^a is 0. The estimate x^a minimizes the statistical quadratic estimation error on any component of the state vector x.

Equations [1.2] and [1.3] are invariant in any invertible linear change of coordinates in either state or data space. This means, for instance, that an observed temperature profile in the data can be transformed into a geopotential profile, through the linear hydrostatic equation, without affecting the final analysis. It also means that the analysis of the horizontal wind field (if spatially discretized) can be performed either in terms of vector coordinates or in terms of divergence and vorticity.

The invariance also means that the analysis is independent of the possible choice of a norm or scalar product in either state or data space (it is not necessary, for instance, to decide for performing the analysis if a 1 $m.s^{-1}$ – error on the wind is larger or smaller than a 1 K – error on the temperature).

In the Bayesian perspective considered in this chapter, we can note that a sample of independent realizations of the conditional probability distribution $\mathcal{N}(x^a, \mathbf{P}^a)$ can easily be obtained. Perturb the data vector z according to:

$$z \rightarrow z' \equiv z + \zeta' \qquad [1.4]$$

where ζ' is distributed according to the data error distribution $\mathcal{N}(0, \mathbf{S})$. It is easily seen that x'^a, defined by performing the analysis on the perturbed data z', viz:

$$x'^a = (\mathbf{\Gamma}^T\mathbf{S}^{-1}\mathbf{\Gamma})^{-1}\mathbf{\Gamma}^T\mathbf{S}^{-1}z' \qquad [1.5]$$

is distributed according to $\mathcal{N}(x^a, \mathbf{P}^a)$.

If the determinacy condition rank$\mathbf{\Gamma} = m$ is verified, it is always possible to decompose the data vector z as follows:

$$x^b = x + \zeta^b \qquad [1.6]$$

$$y = \mathbf{H}x + \epsilon \qquad [1.7]$$

where x^b belongs to state space and has therefore dimension n, while y has dimension $p = m - n$. In addition, the decomposition can be performed in such a way that the "errors" ζ^b and ϵ are statistically uncorrelated, viz.

$$E(\zeta^b\epsilon^T) = 0 \qquad [1.8]$$

The vector x^b will be called a *background* estimate of the state vector x, while the vector y will be called an *observation* vector, belonging to *observation space*, and associated with a linear observation operator $\mathbf{H}$ acting from state into observation space. This vocabulary corresponds to a common situation in geophysics, where the available data consist of a prior estimate x^b (for instance, the output of a recent forecast) and of a set of recent observations y. However, the representation [1.6]–[1.8] is of course absolutely general, and is not restricted to the particular situation of meteorological prediction.

Defining $\mathbf{P}^b \equiv E(\zeta^b\zeta^{bT})$ and $\mathbf{R} \equiv E(\epsilon\epsilon^T)$, equations [1.2] and [1.3] transform under conditions ([1.6]–[1.8]) into either one of the two following equivalent sets of equations:

$$x^a = x^b + \mathbf{P}^b\mathbf{H}^T(\mathbf{H}\mathbf{P}^b\mathbf{H}^T + \mathbf{R})^{-1}(y - \mathbf{H}x^b) \quad [1.9]$$

$$\mathbf{P}^a = \mathbf{P}^b - \mathbf{P}^b\mathbf{H}^T(\mathbf{H}\mathbf{P}^b\mathbf{H}^T + \mathbf{R})^{-1}\mathbf{H}\mathbf{P}^b \quad [1.10]$$

$$x^a = x^b + \mathbf{P}^a\mathbf{H}^T\mathbf{R}^{-1}(y - \mathbf{H}x^b) \quad [1.11]$$

$$[\mathbf{P}^a]^{-1} = [\mathbf{P}^b]^{-1} + \mathbf{H}^T\mathbf{R}^{-1}\mathbf{H} \quad [1.12]$$

In equations [1.9] and [1.11], the analysis x^a is expressed as the sum of the background x^b and a correction proportional to the vector $d \equiv y - \mathbf{H}x^b$. The latter is the difference between the observation y and the analogue $\mathbf{H}x^b$ of the observations in the background. It represents the new information contained in the observation with respect to the background. It is called the *innovation vector*. The matrix applied to the innovation, viz.,

$$\mathbf{K} \equiv \mathbf{P}^b\mathbf{H}^T(\mathbf{H}\mathbf{P}^b\mathbf{H}^T + \mathbf{R})^{-1} = \mathbf{P}^a\mathbf{H}^T\mathbf{R}^{-1} \quad [1.13]$$

is called the *gain matrix*.

It is seen in either [1.10] or [1.12] that $\mathbf{P}^b < \mathbf{P}^a$. The difference represents the improvement in the knowledge of the state vector x brought, in addition to the background x^b, by the observation vector y.

1.2.1. *Variational form*

It can be seen that the analysis x^a (equation [1.2]) minimizes the following scalar *objective function*, defined on the state space $\mathcal{S}$:

$$\xi \in \mathcal{S} \rightarrow \mathscr{J}(\xi) \equiv \frac{1}{2}(\mathbf{\Gamma}\xi - z)^T\mathbf{S}^{-1}(\mathbf{\Gamma}\xi - z) \quad [1.14]$$

$\mathscr{J}(\xi)$ measures the difference in the data space between the data vector z and what the data operator $\mathbf{\Gamma}$ would produce if it was applied on ξ. It is seen in addition that the

Hessian $\mathcal{J}'' \equiv \mathbf{\Gamma}^T \mathbf{S}^{-1} \mathbf{\Gamma}$ of $\mathcal{J}$ (matrix of the second derivative) is the inverse of the analysis error covariance matrix $\mathbf{P}^a$:

$$\mathcal{J}'' = [\mathbf{P}^a]^{-1} \qquad [1.15]$$

In the representation [1.6]–[1.7], the objective function $\mathcal{J}(\xi)$ assumes the following form:

$$\mathcal{J}(\xi) = \frac{1}{2}(\xi - x^b)^T [\mathbf{P}^b]^{-1}(\xi - x^b) + \frac{1}{2}(\mathbf{H}\xi - y)^T \mathbf{R}^{-1}(\mathbf{H}\xi - y) \qquad [1.16]$$

One way to compute the analysis x^a is therefore to minimize the objective function in [1.14] or [1.16].

Equations [1.2]–[1.3] or [1.14] (or equivalently equations [1.9]–[1.10], [1.11]–[1.12] or [1.16]) fully solve the problem of Bayesian estimation in the case of linear operators and additive Gaussian errors. They have actually a broader significance. Coming back to equation [1.1], let us assume that we know nothing about the error ζ beyond the fact that it is a random variable with expectation 0 and covariance matrix $\mathbf{S}$. We want to determine the linear combination $z \in \mathcal{D} \rightarrow \mathbf{A}z \in \mathcal{S}$ that minimizes the statistical quadratic error $E[(\mathbf{A}z - x)(\mathbf{A}z - x)^T]$. The answer is $\mathbf{A}z = (\mathbf{\Gamma}^T \mathbf{S}^{-1} \mathbf{\Gamma})^{-1} \mathbf{\Gamma}^T \mathbf{S}^{-1} z$ as in equation [1.2], the associated minimum statistical quadratic error being $(\mathbf{\Gamma}^T \mathbf{S}^{-1} \mathbf{\Gamma})^{-1}$ as in equation [1.3].

The estimate $x^a = \mathbf{A}z$ is called in this context the *Best Linear Unbiased Estimate (BLUE)* of x from z. It is defined, together with the associated estimation error, by the same algebraic equations ([1.2] and [1.3]) as the Bayesian estimate, but it now has no Bayesian significance. The *BLUE* will not necessarily be the conditional expectation of x for given z. Perhaps, more importantly, the matrix $\mathbf{P}^a = (\mathbf{\Gamma}^T \mathbf{S}^{-1} \mathbf{\Gamma})^{-1}$ will not be the conditional covariance matrix of x for given z, but only the error covariance matrix over all possible realizations of z.

The fact that equations [1.2] and [1.3] solve the more general problem of the *BLUE* gives them additional usefulness for the practical solution of estimation problems. These equations are used, independently of a possible strict Bayesian significance, in many situations. Actually, most algorithms that have been used, or are now used, in assimilation of meteorological or oceanographical observations can be described as more or less heuristic and empirical extensions of equations [1.2] and [1.3] or [1.9] and [1.10] to moderately non-Gaussian and/or nonlinear situations. Many variants exist as to the choice of the operators $\mathbf{\Gamma}$ or $\mathbf{H}$ (which can be nonlinear), the associated covariance matrices $\mathbf{S}$, $\mathbf{P}^b$ or $\mathbf{R}$, and as to the numerical algorithms used for performing the required computations. Algorithms such that OI,

3D and *4D Variational Assimilation*, *Kalman Filter* and *Kalman Smoother* and their various extensions, which will be further described, can all be described as fundamentally emanating from equations [1.2] and [1.3]. The only exception is *Particle Filters*, which are totally independent of any linear or Gaussian hypothesis, and will be described in section 1.5.

Concerning nonlinearity, it can easily be introduced, if it is not too strong, into the above equations. Assume, for instance, that the available data are in the following form, analogous to equations [1.6]–[1.7]:

$$x^b = x + \zeta^b \qquad [1.17]$$

$$y = H(x) + \epsilon \qquad [1.18]$$

H being now a nonlinear operator from state space into observation space. Let us assume that the difference $x - x^b$ is small enough so that the vector $d' \equiv y - H(x^b)$, which is analogous to the innovation vector d introduced above, can be approximated as:

$$d' = y - H(x^b) = H(x) - H(x^b) + \epsilon \approx \mathbf{H}'(x^b)(x - x^b) + \epsilon \qquad [1.19]$$

where $\mathbf{H}'(x^b)$ is the Jacobian matrix (matrix of partial derivatives) of the components of $H(x)$ with respect to the components of x, taken at point x^b. Equations [1.17] and [1.19] then define a linear estimation problem of form [1.6–1.7], with unknown $x - x^b$.

Similarly, if the data operator $\mathbf{\Gamma}$ in equation [1.1] is replaced by a nonlinear operator $\mathbf{\Gamma}$, the latter can be introduced in place of $\mathbf{\Gamma}$ in the objective function (equation [1.14]). Minimization of the new objective function can be expected to lead to a useful estimate of the state vector x, even if the minimum has no precise Bayesian significance. This is largely confirmed by the long experience of variational assimilation, described in the following two sections.

1.3. Optimal interpolation – three-dimensional variational assimilation

In the context of numerical weather prediction, where new forecasts are regularly performed, while new observations are permanently acquired, a natural way to define the initial conditions of a new forecast is to combine the result of the latest forecast with the observations that have been acquired in the meantime. That is exactly what equations [1.9]–[1.10] (or equivalently [1.11]–[1.12]) can achieve, the background x^b being the latest forecast, and the vector y consisting of the new observations. One difficulty here is to define the covariance matrix $\mathbf{P}^b$ of the background error. That was done first by assuming that $\mathbf{P}^b$ is independent of the state of the flow and may depend only on the season of the year and/or the geographical area. This was implemented early in numerical weather prediction, under the name of OI. Before we describe how $\mathbf{P}^b$ was defined in that context in more detail, it is appropriate to make some comments on the significance of equation [1.9].

The vector:

$$\mu \equiv (\mathbf{HP}^b\mathbf{H}^T + \mathbf{R})^{-1}(y - \mathbf{H}x^b) \qquad [1.20]$$

in the right-hand side of equation [1.9], which has dimension $p = m - n$ and components (μ_j) $(j = 1, \ldots, p)$, depends only on the data vectors x^b and y, and is common to all components of the analyzed state vector x^a. It is multiplied in equation [1.9] by the $(n \times p)$-matrix $\mathbf{P}^b\mathbf{H}^T$, whose columns c_j are the covariance vectors of the p observations with the state vector x. Equation [1.9] can therefore equivalently be written as:

$$x^a = x^b + \sum_{j=1,\ldots,p} \mu_j c_j \qquad [1.21]$$

which shows that the modification made by the analysis on the background x^b is a linear combination of p vectors in state space, called the *representers*, which are individually associated with the observations. The representer c_j associated with the component y_j of the vector y defines how y_j influences, through the corresponding covariances, the variables in state space. It is the representers that propagate, in the state space, the information contained in the observations.

One way to numerically compute the analysis x^a is therefore to solve the p-dimensional linear system which defines the vector μ in equation [1.20], and then to perform the multiplications implied by equation [1.21].

In OI, the covariance matrix $\mathbf{P}^b$ has often been defined on the basis of statistical climatological information. One important aspect is the need for enforcing an approximate geostrophic balance between the mass and velocity increments in middle latitudes. Derber and Bouttier (1999) define an approach in which the increments on the geopotential, divergence and vorticity fields are sequentially balanced.

We may often wish to build covariance matrices from explicit analytical expressions. However, we must be careful that the obtained matrices are not only symmetric, but also positive definite. That is actually a very strong constraint. Considering, for instance, homogeneous functions of the form C$[|M_1M_2|]$, where

$|M_1 M_2|$ is the distance (e.g. spherical distance) between points M_1 and M_2, the class of functions C which define proper mathematical covariances is restricted (in brief terms, they are the functions whose spectral transform is real and non-negative). This problem is discussed, in particular, by Gaspari and Cohn (1999), who give explicit analytical covariance functions appropriate for geophysical applications.

As concerns numerical implementation, one major step has been to determine the analysis state x^a by minimization of an objective function of form [1.16]. This is called *Three-Dimensional Variational Assimilation* (*3D-Var*) and is achieved through an iterative process, each step of which requires the explicit determination of the gradient (partial derivatives) of the objective function $\mathscr{J}(\xi)$ with respect to ξ. The latter is called in this context the *control variable*. In view of the large numerical dimension of the state vector, the gradient is determined through the *adjoint* approach to be discussed in more detail in section 1.4.2. The variational approach avoids the need for explicit determination of the gain matrix. It also allows us, as already mentioned, to use nonlinear operators $H(x)$, as for instance the operators associated with a number of satellite observations. This has turned out to be extremely useful and efficient.

3D-Var has been used, and is still used, in operational applications, as well as for research purposes. It was first developed at the ECMWF, where it was actually used as a prior step before developing the more general *Four-Dimensional Variational Assimilation* (*4D-Var*) to be described in the next section. It is to be noted that the expression 3D-Var has come to denote, in the established vocabulary of assimilation of meteorological observation, not any process that is variational and implemented in three dimensions at a given time, but more precisely minimization, at a given time, of an objective function of form [1.16] (with a possibly nonlinear observation operator $H(x)$), where the background error covariance matrix $\mathbf{P}^b$ is independent of the current state of the flow; in particular, of the instability processes that may be at play, or may have been recently at play, in the flow.

1.4. Taking the dynamics of the flow into account

Instability processes, which fundamentally result from temperature gradients (which result themselves from differential heating), are ubiquitous in the atmosphere. Everything else being the same, the uncertainty in the state of the flow at a given time is likely to be larger in those components of the flow that have been subject to recent instability, and it is certainly preferable to take the corresponding instability processes into account in the assimilation. Figure 1.1 shows, for a given meteorological situation, the temporal evolution of the error on the state of the flow over a simulated 24-h forecast. The figure has been obtained by repeated integration of a simple, but basically realistic, model of the atmospheric circulation. The quantity represented is the spatial correlation of the error between a given fixed point (45N, 35W) and surrounding points. It is seen that as the forecast range increases from 0 to 6 and 24 h (respective panels from (a) to (c)), the spatial extent of the correlation pattern increases, while its shape is significantly modified. In addition, the amplitude of the uncertainty increases as well as a result of the sensitivity of the forecast to the initial condition (not shown). The important fact here is that the evolution visible in Figure 1.1 will be different in another meteorological situation, and that it is therefore highly desirable to take into account the specific instability properties of the current situation in each new assimilation.

Kalman Filtering and Variational Assimilation, which we are going to describe now, are two broad classes of algorithms that are widely used in assimilation of meteorological and oceanographical observations. They allow taking into account, in a mathematically consistent way, the dynamical evolution of the flow together with the associated uncertainty. Although these algorithms significantly differ in their numerical implementation, they both derive from the basic estimate (equations [1.2] and [1.3]) and are equivalent in the linear case. It is only in nonlinear situations that they can lead to different results.

We explicitly introduce the temporal evolution of the system under observation. The state space $\mathcal{S}$, still with dimension n, is now the space in which the state vector, denoted x_k at time k, evolves in time. The evolution is assumed to be governed by the equation:

$$x_{k+1} = \mathbf{M}_k x_k + \eta_k \qquad [1.22]$$

where the linear operator $\mathbf{M}_k$ is a known *model operator*, which expresses our knowledge of the dynamics governing the evolution of the flow. The last term η_k is called the *model error*. It corresponds to the fact that the model operator $\mathbf{M}_k$ will not in general be exact, and that the real state x_{k+1} at time $k+1$ will not in general be obtained by applying that operator to the real state vector x_k at time k (assuming the latter to be known). The model error is the error in the model representation of the dynamics of the flow. It should be stressed that it is not in general the error in a forecast produced by the model. The error in a forecast results, not only from the model error, but also from the error in the initial conditions of the forecast.

The model error η_k will be assumed to be random with expectation 0, to be uncorrelated in time, and to have covariance matrix $\mathbf{Q}_k$ at time k. This is expressed by the equation:

$$E(\eta_k \eta_{k'}^T) = \delta_{kk'} \mathbf{Q}_k \qquad [1.23]$$

where $\delta_{kk'}$ is the Kronecker symbol ($\delta_{kk'} = 1$ if $k = k'$, and $\delta_{kk'} = 0$ if $k \neq k'$).

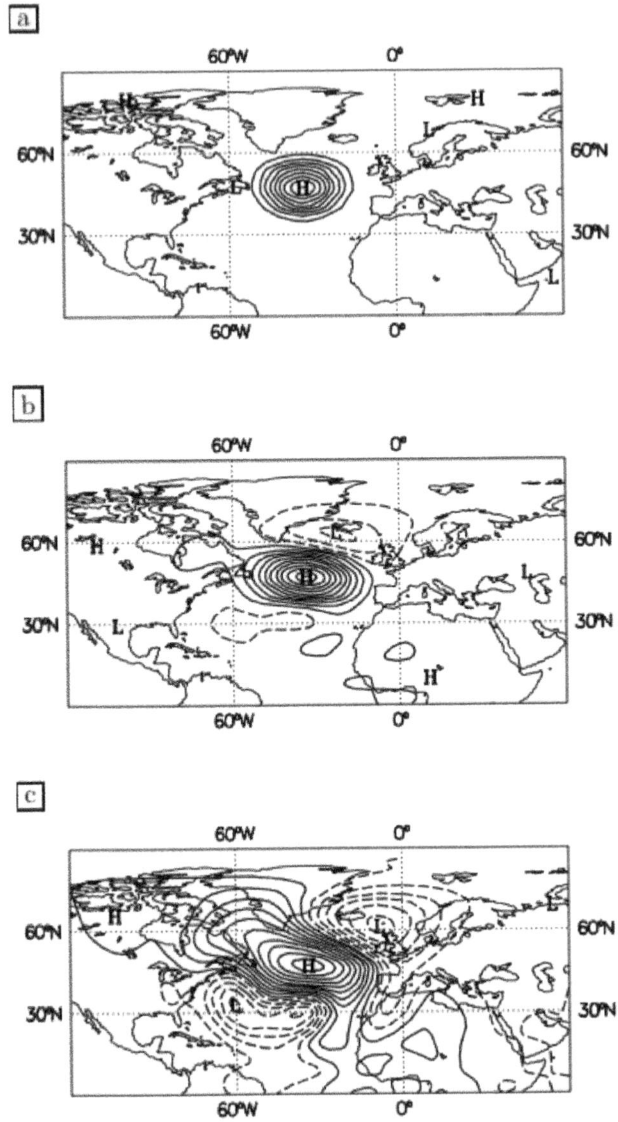

Figure 1.1. *Spatial correlation (between point 45N, 35W and surrounding points) of the error in a simulated numerical forecast at times 0, 6 and 24 h (from (a) to (c)). The field under consideration is the geopotential of the 500 hPa isobaric surface. The initial correlation (a) has been chosen as isotropic and everywhere positive. It is seen that, as the forecast proceeds, the correlation loses its isotropy while its spatial extension increases, and that negative values of correlation appear (dashed contours) (credit: F. Bouttier (Bouttier 1994))*

We assume that observations are available at successive times $k = 0, \ldots, K$, in the form:

$$y_k = \mathbf{H}_k x_k + \epsilon_k \qquad [1.24]$$

where the observation vector y_k has dimension p, $\mathbf{H}_k$ is a known linear operator from the state space into the observational space, represented by a $p \times n$ matrix. ϵ_k is a random observational error with expectation 0, is uncorrelated in time, and has covariance matrix $\mathbf{R}_k$ at time k, viz.,

$$E(\epsilon_k \epsilon_{k'}^T) = \delta_{kk'} \mathbf{R}_k \qquad [1.25]$$

(no explicit dependence of the dimension p of the observation vector with time is considered, but such a dependence could easily be introduced if necessary).

We assume that a background estimate x_0^b is available at the initial time $k = 0$, viz.,

$$x_0^b = x_0 + \zeta_0^b \qquad [1.26]$$

with error covariance matrix:

$$E(\zeta_0^b \zeta_0^{bT}) = \mathbf{P}_0^b \qquad [1.27]$$

Finally, we assume that the model errors η_k, the observation errors ϵ_k and the background error ζ_0^b are all mutually uncorrelated.

In these conditions, equations [1.22] to [1.27] define a linear estimation problem, which we will denote $\mathcal{P}$, of the general form (equation [1.1]), the unknown being now the consequence of system states $(x_0^T, \ldots, x_k^T, \ldots, x_K^T)^T$. The presence of the background (equations [1.26] and [1.27]), together with equation [1.22], ensures that the underlying determinacy condition is verified. The corresponding objective function (equation [1.16]) reads:

$$\begin{aligned}
&(\xi_0^T, \ldots, \xi_k^T, \ldots, \xi_K^T)^T \in \mathcal{S}^{K+1} \rightarrow \\
&\mathscr{J}_P(\xi_0, \ldots, \xi_k, \ldots, \xi_K) \equiv \frac{1}{2}(\xi_0 - x_0^b)^T [\mathbf{P}_0^b]^{-1} (\xi_0 - x_0^b) \\
&\qquad + \frac{1}{2} \sum_{k=0,\ldots,K} [y_k - \mathbf{H}_k \xi_k]^T \mathbf{R}_k^{-1} [y_k - \mathbf{H}_k \xi_k] \qquad [1.28] \\
&\qquad + \frac{1}{2} \sum_{k=0,\ldots,K-1} [\xi_{k+1} - \mathbf{M}_k \xi_k]^T \mathbf{Q}_k^{-1} [\xi_{k+1} - \mathbf{M}_k \xi_k]
\end{aligned}$$

where the term on the first line of the right-hand side measures the misfit of the unknown to the background x_0^b (equations [1.26] and [1.27]), the summation on the second line measures the misfit to the observations y_k (equations [1.24] and [1.25]) and the summation on the third line measures the misfit to the model equation (equations [1.22] and [1.23]).

1.4.1. *The Kalman Filter*

The Kalman Filter (Kalman 1960) is an algorithm for solving the estimation problem (equations [1.22] to [1.27]) recursively over time. It starts from the initial background x_0^b and the associated error covariance matrix $\mathbf{P}_0^b$. Assume a background estimate x_k^b has been obtained at a later time k from data (initial background, model equation and observations) at previous times, together with an error covariance matrix $\mathbf{P}_k^b$. The observation vector y_k is then used, through equations [1.9]–[1.10], to update these background estimates and to obtain a new, analyzed estimate x_k^a together with its error covariance matrix $\mathbf{P}_k^a$, viz.,

$$x_k^a = x_k^b + \mathbf{P}_k^b\mathbf{H}_k^T(\mathbf{H}_k\mathbf{P}_k^b\mathbf{H}_k^T + \mathbf{R}_k)^{-1}(y_k - \mathbf{H}_k x_k^b) \quad [1.29]$$

$$\mathbf{P}_k^a = \mathbf{P}_k^b - \mathbf{P}_k^b\mathbf{H}_k^T(\mathbf{H}_k\mathbf{P}_k^b\mathbf{H}_k^T + \mathbf{R}_k)^{-1}\mathbf{H}_k\mathbf{P}_k^b \quad [1.30]$$

Following this *analysis step*, a *forecast step* is performed to time $k+1$, leading to:

$$x_{k+1}^b = \mathbf{M}_k x_k^a \quad [1.31]$$

$$\mathbf{P}_{k+1}^b = \mathbf{M}_k\mathbf{P}_k^a\mathbf{M}_k^T + \mathbf{Q}_k \quad [1.32]$$

The background x_{k+1}^b is obtained in equation [1.31] by applying the model equation (equation [1.22]) in its deterministic form to the analysis x_k^a. As for the corresponding estimation error covariance matrix $\mathbf{P}_{k+1}^b$, given by equation [1.32], it is obtained from equations [1.22] and [1.23], together with the hypothesis of temporal decorrelation of the errors. The term $\mathbf{M}_k\mathbf{P}_k^a\mathbf{M}_k^T$ on the right-hand-side corresponds to the error at time $k+1$ that results from the analysis error at time k, after it has been transported by the model $\mathbf{M}_k$. The term $\mathbf{Q}_k$ represents the effect of the model error η_k between times k and $k+1$ (equations [1.22] and [1.23]). Since the model error is assumed to be decorrelated from all previous errors, the corresponding covariance matrices simply add up.

The Kalman filter, starting from the initial conditions x_0^b and $\mathbf{P}_0^b$, thus consists of an alternative sequence of analysis and forecast steps of the respective forms [1.29]–[1.30] and [1.31[–[1.32]. At any time k, it solves the estimation problem $\mathcal{P}$ stated above. More precisely, if all errors in the data are globally Gaussian, the Gaussian probability distribution $\mathcal{N}(x_k^b, \mathbf{P}_k^b)$ [respectively $\mathcal{N}(x_k^a, \mathbf{P}_k^a)$] is the probability distribution for the real state x_k, conditioned by all data prior to (respectively up to) time k. In the case that the errors are not Gaussian, but the model and observation operators are still linear, a similar statement applies; the filter does not producing conditional probability distributions for the state x_k, but *BLUEs* for that state.

The idea of applying the Kalman Filter to assimilation of meteorological observations came at a very early stage, the first contribution being apparently by Jones (1965), with other early works by Petersen (1968); Talagrand and Miyakoda (1971) and Ghil et al. (1981).

Equation [1.32] achieves what has been identified above as a desirable aspect of assimilation, namely describing the temporal evolution the uncertainty on the state of the flow. Its cost is, however, prohibitive in any practical situation. Taken literally, it requires $2n$ multiplications by the matrix $\mathbf{M}_k$, that is, $2n$ integrations of the model between times k and $k+1$. That would be totally impossible in real time with a realistic model. Two approaches have been defined to circumvent that difficulty: *reduced order filters,* on the one hand, in which the uncertainty is assumed to be concentrated in a low-dimensional subspace of the state space, and *ensemble filters.* on the other hand, in which the uncertainty is represented by a finite number of points (model states) in state space.

1.4.1.1. *Reduced order filters*

There are several variants of reduced order filters. We succinctly describe here the *reduced rank square root Kalman filter* (RRSQRT), introduced by Verlaan and Heemink (1997, 2001). In that approach, the analysis is performed in the subspace spanned by $r \ll n$ eigenvectors of the background error covariance matrix $\mathbf{P}^b$ (we drop the time index k). More precisely, an $n \times n$ covariance matrix $\mathbf{P}$ can be written as:

$$\mathbf{P} = \mathbf{E}\mathbf{E}^T \qquad [1.33]$$

where the column vectors of the $(n \times n)$-matrix $\mathbf{E}$ are the (orthogonal) eigenvectors of $\mathbf{P}$ (the modulus of each vector being the square root of the associated positive eigenvalue). The RRSQRT consists of writing the matrix $\mathbf{P}^b$ in the form:

$$\mathbf{P}^b \approx \mathbf{E}^b\mathbf{E}^{bT} \qquad [1.34]$$

where $\mathbf{E}^b$ is an $(n \times r)$-matrix, the columns of which are the r dominant eigenvectors of $\mathbf{P}^b$ (or some appropriate approximation thereof). Setting $\mathbf{\Psi} \equiv (\mathbf{HE^b})^\mathbf{T}$, the analysis gain matrix (equation [1.13]) becomes:

$$\mathbf{P}^b\mathbf{H}^T(\mathbf{H}\mathbf{P}^b\mathbf{H}^T + \mathbf{R})^{-1} \approx \mathbf{E}^b\mathbf{\Psi}(\mathbf{\Psi}^T\mathbf{\Psi} + \mathbf{R})^{-1} \qquad [1.35]$$

while the analysis error covariance matrix $\mathbf{P}^a$ (equation [1.10]) becomes:

$$\mathbf{P}^a \approx \mathbf{E}^a \mathbf{E}^{aT} \qquad [1.36]$$

with:

$$\mathbf{E}^a = \mathbf{E}^b [\mathbf{I}_r - \mathbf{\Psi}(\mathbf{\Psi}^T \mathbf{\Psi} + \mathbf{R})^{-1} \mathbf{\Psi}^T]^{1/2} \qquad [1.37]$$

The matrix $\mathbf{\Psi}$ has dimension $p \times r$, so that, if the number p of observations is small, formulas [1.35] and [1.36]–[1.37] allow us to compute the (approximate) analysis x^a and the matrix $\mathbf{E}^a$ at a low cost.

In the forecast step, the r column vectors of $\mathbf{E}^a$ are evolved with the model (equation 1.22). A model error, if it is to be introduced, is not introduced in RRSQRT by an explicit perturbation term as in equation [1.22]. It can be introduced by decreasing the number of column vectors to be integrated, and replacing them by vectors meant to represent the dominant directions of the model error.

The main advantage of RRSQRT, provided that the number r of eigenvectors required to describe the dynamics (equation [1.22]) to a good degree of accuracy is small, is its low cost. Another advantage is the definition of the covariance matrices by their square root, as in equation [1.36]. This ensures the necessary positive definiteness of the covariance matrices, which can be lost through poor numerical conditioning in other algorithms. On the other hand, the dominance of the column vectors in $\mathbf{E}^b$ may be lost in the temporal integration of the model, thus degrading the accuracy of the filter. New, dominant eigenvectors may have to then be redefined.

Other forms of somewhat similar reduced order filters have been defined and studied, such as the *Singular Evolutive Extended Kalman Filter* (SEEK) (Pham et al. 1998) and its extension to nonlinear dynamics, the *Singular Evolutive Interpolated Kalman Filter* (SEIK) (Pham 2001).

It should be noted that these reduced order filters require the identification of directions of largest uncertainty in state space. That, contrary to what is the case for the Kalman Filter itself, which is norm-independent, requires the use of a norm in that space.

1.4.1.2. *The Ensemble Kalman Filter*

The EnKF (originally proposed for oceanographical applications by Evensen (1994), see also Evensen (2009)) is a Monte Carlo type method, in which the uncertainty on the state of the flow is at any time described by a finite ensemble of L

points in the state space (i.e. of state vectors). This ensemble is evolved through successive alternate analysis and forecast steps, similar to what is done in the standard Kalman Filter.

Let us assume that an ensemble of L background estimates $\{x_l^b\}$ $(l = 1, \ldots, L)$ is available at time k, at which an observation $y = \mathbf{H}x + \epsilon$ is available (we again drop the time index). The purpose is to update each background estimate x_l^b through a formula of form [1.9] in order to produce an analysis estimate x_l^a. To this end, an estimate of the background error covariance matrix $\mathbf{P}^b$ must be available. One possible choice is to take the sample covariance matrix of the ensemble $\{x_l^b\}$, viz.,

$$\mathbf{P}^b_{En} \equiv \frac{1}{L-1} \sum_{l=1,\ldots,L} (x_l^b - \bar{x}^b)(x_l^b - \bar{x}^b)^T \qquad [1.38]$$

where $\bar{x}^b$ is the average $\bar{x}^b \equiv \frac{1}{L} \sum_{l=1,\ldots,L} x_l^b$. In addition, the observation vector y is transformed through perturbations of the form:

$$y \rightarrow y_l = y - \epsilon_l \quad l = 1, \ldots, L \qquad [1.39]$$

into an ensemble of L approximate observations. The perturbations ϵ_l are independent realizations of the probability distribution of the observation error ϵ. The y_l's are meant to sample the uncertainty on the observation y. The couples $\{x_l^b, y_l\}$ are then combined together through the following formula, of form [1.9]:

$$x_l^a = x_l^b + \mathbf{P}^b_{En}\mathbf{H}^T(\mathbf{H}\mathbf{P}^b_{En}\mathbf{H}^T + \mathbf{R})^{-1}(y_l - \mathbf{H}x_l^b) \quad l = 1, \ldots, L \qquad [1.40]$$

This produces an ensemble of L analyzed states $\{x_l^a\}$, from which we can compute, for example, a mean analysis or an analysis error covariance matrix.

In the forecast step of the filter, the model (equation [1.22]) is integrated from the L analyzed states $\{x_l^a\}$. If a model error η_k is present, it can be added as a random perturbation in the course of each integration. And if, contrary to what is the case in equation [1.22], the model itself is nonlinear, the L integrations can also be performed with that model, thus eliminating the need for any linearity hypothesis in the dynamics of the system.

Equation [1.40] requires a linear observation operator $\mathbf{H}$ in the gain matrix. If the observation operator, denoted now H, is nonlinear, one solution is to replace it by its local Jacobian $\mathbf{H}'$ (see equation [1.19]). Another solution is to note that the matrices $\mathbf{P}^b\mathbf{H}^T$ and $\mathbf{H}\mathbf{P}^b\mathbf{H}^T + \mathbf{R}$ in equation [1.9] are actually the covariance matrices $E[(x - \bar{x}^b)d^T]$ and $E(dd^T)$. These can be approximated here as sample covariances from the ensembles $\{x_l^b - \bar{x}^b\}$ and $\{y_l - H(x_l^b)\}$ $(l = 1, \ldots, L)$ (if the latter solution is used, it avoids computation of the matrix $\mathbf{P}_{En}^b$ in equation [1.38]).

Experience shows that EnKF leads to very good results, even with ensembles of relatively small size, on the order of a few tens or at most 100–200, much smaller in any case than the dimension of the state space. Through the temporal integration of the ensemble, EnKF takes into account the temporal evolution of the uncertainty of the flow. This has a significant impact on its performance. As an example, Meng and Zhang (2008) have made a methodical comparison, over a long period, of an EnKF and a 3D-Var system, using the same model and the same observations. They have found a systematic advantage for the former. This advantage is due to the temporal integration of the ensemble.

Practical implementation of EnKF nevertheless raises a number of difficulties. The relatively small dimension of the ensembles, if it allows us to perform the required computations in the first place, has some disadvantages. It is observed that the covariances defined by the sample matrix $\mathbf{P}_{En}^b$ (equation [1.38]) show unrealistic large values at long spatial distances. This is due to undersampling by a relatively small ensemble. In addition, the matrix $\mathbf{P}_{En}^b$ has rank $L - 1$ at most, so that the increments defined by equation [1.40] are restricted from the start to a space of relatively small dimension, which does not allow us to properly sample the uncertainty on the analysis. Both these aspects are dealt with heuristically by an appropriate method of *localization*. The simplest localization method consists of modifying background covariances so that they smoothly decrease to zero at long distances. This can be achieved by replacing $\mathbf{P}_{En}^b$ with the matrix $\mathbf{P}_{Loc}^b$ defined by:

$$\mathbf{P}_{Loc}^b \equiv \mathbf{L} \circ \mathbf{P}_{En}^b \qquad [1.41]$$

In this expression, $\mathbf{L}$ is a symmetric isotropic definite positive matrix whose entries are close to unity at short distance and smoothly decrease with distance to become equal to zero beyond a distance d. The notation $\circ$ denotes Schur multiplication, or entry-wise multiplication of the two matrices. The matrix $\mathbf{P}_{Loc}^b$ thus defined is symmetric definite positive, which ensures that it is a proper covariance matrix, and achieves the intended purpose of small (and even zero) covariances at long distances. The choice of an appropriate matrix $\mathbf{L}$ can be based on the compactly supported covariance functions defined by Gaspari and Cohn (1999).

In addition, the localized matrix $\mathbf{P}_{Loc}^b$ is usually of full rank, which avoids the a priori concentration of the analyzed sample on a subspace that may be far from the observations.

Another difficulty with EnKF is that analyzed ensembles of small dimension tend to collapse and lose dispersion. Too small dispersion can lead to divergence of the filter, which will "trust" the background ensemble too much, and gradually diverge from the observations. The collapse is due, at least in part, to the nonlinear dependence of the gain matrix (equation [1.13]) with respect to the background error covariance matrix $\mathbf{P}^b$. This creates a systematic curvature bias, which decreases the dispersion of the ensembles. The usual procedure to avoid the collapse is to inflate the analyzed ensemble by an appropriate coefficient.

EnKF is widely used in many practical applications. Many variants have been defined. Most of these include both localization and inflation of ensembles, with coefficients largely determined through problem-dependent trial-and-error experimentation. One particular approach has been defined by Houtekamer and Mitchell (2001), who use two ensembles, the covariance matrix used for updating one ensemble being obtained from the other ensemble. A description of many of those variants can be found in Chapter 6 of Asch et al. (2016). Much work has been done also on the theoretical aspects of EnKF. For instance, Bocquet (2011) and Bocquet et al. (2015) have defined a filter which takes into account the fact that the background ensemble $\{x_l^b\}$, being a finite size sample, does not exactly represent the probability distribution of the background error. This approach must avoid the need for inflation.

In the Bayesian perspective taken here, it is interesting to know whether EnKF achieves Bayesianity. The standard Kalman Filter (equations [1.29]–[1.32]) achieves Bayesianity in the linear and Gaussian case under the hypothesis of decorrelation of errors in time. Under the same hypothesis, assuming the background ensemble $\{x_l^b\}$ to be a sample of independent realizations of the underlying conditional probability distribution $\mathcal{N}(x^b, \mathbf{P}^b)$, can a similar statement be made about the analysis $\{x_l^a\}$? If the background error covariance matrix used in the update (equation [1.40]) was equal to $\mathbf{P}^b$, the remark made about equations [1.4]–[1.5] would show that the answer is positive. Now, as just stressed above, the matrix $\mathbf{P}^b_{En}$ in equation [1.40] is not equal to $\mathbf{P}^b$ (nor would any other matrix defined from the finite ensemble $\{x_l^b\}$, such as $\mathbf{P}^b_{Loc}$ in equation [1.41]), so that the answer will in general be negative. However, we can note that the forecast step of EnKF, if started from a possible Bayesian analysis ensemble, would preserve Bayesianity of the ensuing background ensemble.

Le Gland et al. (2011) have studied the asymptotic properties of the EnKF when the dimension L of the ensemble grows to infinity. They have found that the distribution of the ensemble tends to an asymptotic limit, but that it is only in the linear case that the limit coincides with the Bayesian probability distribution.

The Kalman Filter, as described above, either in its standard form (equations [1.29]–[1.32]) or in its ensemble form, carries information forward in time only. It does not propagate the information backward in time (that is not necessary if assimilation is used only for defining initial conditions of forecasts, but can be useful for a posteriori assimilation of observations, as done in climate reanalyses). Kalman Filter is not capable to take into account temporal correlation between errors (indeed, temporal decorrelation is necessary for the optimality of the standard filter). Such temporal correlation certainly exists in reality. Concerning observation errors, observations performed at successive times by the same instrument (for instance, a satellite-borne radiometer) are certainly correlated. Also, as concerns model errors, they are certainly correlated, at least over short periods of time.

The algorithms that are capable of time-backward propagation of the information contained in the observations are called *smoothers*. One obvious approach for that is of course to consider a four-dimensional problem, using as data vector the set of all data (observations and model) that are available over the intended assimilation window, and taking as unknown the sequence of all model states over that period. That is what minimization of the objective function (equation [1.28]) would achieve in the linear case (in the absence of temporal correlation if we restrict ourself to the expression [1.28]). We will describe how variational assimilation, based on the minimization of an objective function integrated over time, is implemented in operational weather prediction in the next section. However, we mention that a large number of smoothers, some of which are based on a sequential approach similar to Kalman Filtering, have been developed and studied by many authors. All these algorithms propagate the information backward in time, while some are capable of taking temporal correlation into account. We refer the reader interested by a more complete description of smoothers to section 6.8 of Asch et al. (2016) as well as to Cosme (2015).

1.4.2. *Four-dimensional variational assimilation*

1.4.2.1. *Strong constraint variational assimilation*

As already said, equations [1.22]–[1.27] define a linear estimation problem of the general form (equation [1.1]). However, before considering it as such, we consider a different nonlinear problem, which was at the origin of four-dimensional variational assimilation. We consider the dynamical model to be nonlinear, but exact, so that equation [1.22] is changed into:

$$x_{k+1} = M_k(x_k) \qquad [1.42]$$

where M_k is now a nonlinear operator. A solution of equation [1.42] is entirely determined by the specification of x_0 at initial time 0. Similarly, we assume a nonlinear operator H_k in equations [1.24]–[1.25]. Except for that, the observation errors in equations [1.24]–[1.25] are still assumed to be uncorrelated in time, with error covariance matrix $\mathbf{R}_k$ at time k. A background estimate x_0^b is still assumed to be available at time $k = 0$, with error covariance matrix $\mathbf{P}_0^b$. We then define from equations [1.42], [1.24] and [1.25], the objective function:

$$\xi_0 \in \mathcal{S} \rightarrow \mathscr{J}_S(\xi_0) \equiv \frac{1}{2}(\xi_0 - x_0^b)^T[\mathbf{P}_0^b]^{-1}(\xi_0 - x_0^b) + \frac{1}{2}\sum_{k=0,\ldots,K}[y_k - H_k(\xi_k)]^T\mathbf{R}_k^{-1}[y_k - H_k(\xi_k)] \quad [1.43]$$

subject to $\xi_{k+1} = M_k(\xi_k), \quad k = 0, \ldots, K-1$

For any $\xi_0 \in \mathcal{S}, \mathscr{J}(\xi_0)$ measures the misfit between the solution of equation [1.42] emanating from ξ_0, on the one hand, and the background x_0^b and the observations y_k, on the other hand.

Minimization of $\mathscr{J}(\xi_0)$ is called *strong constraint four-dimensional variational assimilation* (Strong Constraint 4D-Var). The words "Strong Constraint" mean that, in contrast with the "Weak Constraint" to be discussed below, the model equation [1.42] is supposed to be exact, with the consequence that the sequence of analyzed states exactly verify that equation.

The main problem with the numerical minimization of a scalar function such as $\mathscr{J}_S(\xi_0)$ is the dimension of the control variable ξ_0. Minimization must be performed by an iterative gradient process, each step of which requires the explicit knowledge of the gradient of the function to be minimized with respect to the control variable. The cost of explicit computation of the gradient by finite perturbation of the components of the control variable, followed by finite difference, would be prohibitive. The most efficient way to compute the gradient is through the *adjoint* approach. The adjoint approach, which was first introduced in the context of assimilation of observations by Penenko and Obraztsov (1976), was later developed by Lewis and Derber (1985) and Le Dimet and Talagrand (1986). It is based on the following mathematical fact. Consider a mapping $u \rightarrow v = F(u)$ with input u and output v (with any dimensions), and a scalar function $\mathscr{J}(v)$ of the output v. The gradient of the compound function $\mathscr{J}_o F$ with respect to the input u, which we simply denote $\nabla_u \mathscr{J}$, is equal to:

$$\nabla_u \mathscr{J} = \mathbf{F}'^T \nabla_v \mathscr{J} \quad [1.44]$$

where $\nabla_v \mathscr{J}$ is the gradient of $\mathscr{J}$ with respect to the output v, and $\mathbf{F}'$ is the Jacobian of the mapping F, taken at point u. This allows us to proceed, through the transpose (or adjoint) $\mathbf{F}'^T$, from the gradient with respect to the output to the gradient with respect to the input. If the mapping F is a composition of the form $F_N \circ \ldots \circ F_1$, the transpose $\mathbf{F}'^T$ is the numerical product, taken in reverse order, of the transposes of the Jacobians of those components, viz.,

$$\mathbf{F}'^T = \mathbf{F}_1'^T \dots \mathbf{F}_N'^T \quad [1.45]$$

In the case of the objective function (equation [1.43]), the adjoint approach, once the input ξ_0 and the corresponding solution $\xi_k (k = 0, \dots, K)$ of equation [1.42] are known, leads to the following sequence of operations for the computation of the gradient of $\mathscr{J}_S$ with respect to ξ_0:

– Compute at the final time K the vector:

$$\lambda_K = \mathbf{H}'^T_K \mathbf{R}_K^{-1}[H_K(\xi_K) - y_K] \quad [1.46]$$

– Then, proceeding backwards in time, compute for $k = K - 1, \dots, 1$:

$$\lambda_k = \mathbf{M}'^T_k \lambda_{k+1} + \mathbf{H}'^T_k \mathbf{R}_k^{-1}[H_k(\xi_k) - y_k] \quad [1.47]$$

– Finally compute at time $k = 0$:

$$\lambda_0 = \mathbf{M}'^T_0 \lambda_1 + \mathbf{H}'^T_0 \mathbf{R}_0^{-1}[H_0(\xi_0) - y_0] + [\mathbf{P}_0^b]^{-1}(\xi_0 - x_0^b) \quad [1.48]$$

In these equations, the matrices $\mathbf{M}'_k$ $(k = 0, \dots, K - 1)$ and $\mathbf{H}'_k$ $(k = 0, \dots, K)$ are the Jacobians of the respective operators M_k and H_k, taken at point ξ_k. The vectors λ_k $(k = 0, \dots, K)$, which have dimension n, are called the *adjoint variables*. The vector λ_0 is the required gradient of the objective function (equation [1.43]) with respect to the initial condition $u \equiv \xi_0$, viz., $\lambda_0 = \nabla_u \mathscr{J}_S$.

In agreement with equation [1.45], the adjoint computations (equations [1.46]–[1.48]) are performed backwards in time, starting from the final time K. Their numerical cost is of the same order of magnitude as the cost of the direct integration of equation [1.42] (about twice as costly in the case of an atmospheric model, that ratio being basically independent of the dimension n of the state vector). It is also seen that they require the explicit knowledge of the current solution $\{\xi_k\}$ $(k = 0, \dots, K)$ of equation [1.42]. The current solution is used as argument for the observation operators H_k, and also, in the case of nonlinear operators M_k and H_k, as arguments in the Jacobians $\mathbf{M}'_k$ and $\mathbf{H}'_k$. It must therefore be computed before the adjoint computations are started, and then kept in memory (or recomputed in the course of the adjoint computations). That high requirement of memory size is the price to be paid for the low computational cost of the adjoint approach.

As said, the objective function $\mathcal{J}_S$ is minimized through an iterative procedure. Each step of that procedure requires one direct integration of equation [1.42], started from the current value of the control variable ξ_0, followed by one backward integration of the adjoint equations (equations [1.46]–[1.48]). Although costly, that is much more economical than computation of the gradient by explicit perturbation of the control variable.

The impact of 4D-Var, as compared with 3D-Var, is very significant. Thépaut et al. (1993) compare the increments produced at the end of a 24-h assimilation by 3D-Var and 4D-Var. While the former shows the increment structure imposed by the a priori choice of the background error covariance matrix $\mathbf{P}^b$ (equation [1.16]), the latter shows a much more complicated structure, which results from the evolution of the flow over the 24 h of assimilation, and in particular of the instabilities that have occurred over that period.

It is the adjoint approach that has made variational assimilation possible in the first place. Strong constraint variational assimilation using the adjoint approach was first introduced in operational numerical weather prediction by ECMWF in 1997 with an assimilation window of 6 h. It led to significant improvements in the quality of the ensuing forecasts in comparison with the previous 3D-Var (see, for example, Klinker 2000; Rabier 2000).

Contrary to Kalman Filtering, variational assimilation does not compute an explicit estimate of the uncertainty on the state of the flow, either in the final analysis or at any other step of the process. However, the Jacobian $\mathbf{M}'_k$ describes to first order the evolution of the difference between two close solutions of the model equation [1.42]. The evolution of the uncertainty on the state of the flow is thus represented, not explicitly, but in a way that allows us to minimize the objective function $\mathcal{J}_S$ (that would produce in the linear case, the same estimate as Kalman Filter at the final time K). Another basic difference is that variational assimilation, in contrast to standard sequential Kalman filter, is smoother in that it adjusts the assimilating model to the whole set of observations available over the assimilation window, and propagates the information contained in the observations both forward and backward in time. This is not important when assimilation is used for defining the initial conditions of a forecast, but can be important for other uses, for instance, for reanalysis of past observations where it is preferable to use all available observations, performed either before or after analysis time.

As for the Kalman filter, approximations have been made for reducing the cost of the computations. These approximations essentially consist of simplifying the assimilating model (equation [1.42]) together with its adjoint (exact consistency between the assimilating model and its adjoint is necessary for the success of the minimization). This is usually achieved by decreasing the resolution of the model, which means that the minimization is performed only in the large scales of the flow, the smaller scales not being modified. In the *incremental approach* described by Courtier et al. (1994), the approximations that are made vary in the successive iterations of the minimization process. The whole procedure is defined in such a way that, owing to appropriate successive linear approximations, the global minimization consists of a sequence of minimizations of more easily minimized exactly quadratic objective functions.

On the other hand, Variational Assimilation, as described above, does not allow cycling for one assimilation window to the next. As described above, it has to be restarted from scratch for each new window. Purely sequential assimilation such as the Kalman Filter, on the contrary, allows continuous introduction of new observations. However, the method of *Ensemble of Data Assimilations*, described in section 1.4.3, allows for carrying partial information from one window of Variational Assimilation to the next.

Strong constraint Variational Assimilation and EnKF are the two main algorithms that are used for assimilation in operational large-dimension numerical weather prediction. No significant difference seems to have been observed between the results they produce. A clean comparison experiment (same model, same observations, same assumed errors) was made by Buehner et al. (2010a, 2010b) who found that, for the same total computational cost, both approaches led to results which, although different in some aspects, were of similar global quality.

1.4.2.2. *Weak Constraint Variational Assimilation*

Weak Constraint Variational Assimilation takes into account the presence of errors in the numerical model, as in equation [1.22]. It is seen from equations [1.28] and [1.43] that, in the case of nonlinear model or observation operators, the corresponding objective function to be minimized reads:

$$
\begin{aligned}
&(\xi_0^T, \ldots, \xi_k^T, \ldots, \xi_K^T)^T \in \mathcal{S}^{K+1} \rightarrow \\
&\mathcal{J}_w(\xi_0, \ldots, \xi_k, \ldots, \xi_K) \equiv \frac{1}{2}(\xi_0 - x_0^b)^T [\mathbf{P}_0^b]^{-1} (\xi_0 - x_0^b) \\
&\qquad + \frac{1}{2} \sum_{k=0,\ldots,K} [y_k - H_k(\xi_k)]^T \mathbf{R}_k^{-1} [y_k - H_k(\xi_k)] \\
&\qquad + \frac{1}{2} \sum_{k=0,\ldots,K-1} [\xi_{k+1} - M_k(\xi_k)]^T \mathbf{Q}_k^{-1} \\
&\qquad \times [\xi_{k+1} - M_k(\xi_k)]
\end{aligned} \qquad [1.49]
$$

which differs from equation [1.43] in that the control variable now consists of the whole sequence of model states $\{\xi_k\}_{k=0,\dots,K}$, and that additional terms, measuring the misfit of that sequence to the dynamical equation [1.42], are present. Equation [1.49] differs from equation [1.28] in that it allows for nonlinear model and observation operators.

Weak Constraint Variational Assimilation has been abundantly studied. One major difficulty is that it requires an explicit specification for the model error covariance matrix $\mathbf{Q}_k$. It is very difficult to quantitatively assess the model error, which can vary significantly with the variables under consideration and the current situation of the flow (not to mention the fact that it can also be correlated in time). On the other hand, Weak Constraint Assimilation has a distinct numerical advantage. The model error term regularizes the objective function $\mathscr{J}_w(\xi_0, \dots, \xi_k, \dots, \xi_K)$ (as any smoothing term actually does) and makes the minimization easier. In spite of that, it does not seem that Weak Constraint Variational Assimilation could easily be implemented in large-scale operational applications, but positive results have been obtained. The operational assimilation system of ECMWF now includes a weak constraint component in the assimilation of stratospheric fields.

1.4.3. *Ensemble methods*

EnKF and Variational Assimilation have been widely used in operational systems for assimilation, with constant improvements but no fundamental changes, for two decades. Research for more powerful methods nevertheless continues, as we are going to describe now.

As said before, it is natural, in the general Bayesian logic that we have described for assimilation, to develop methods which produce an ensemble of state points meant to sample the probability distribution of the state of the observed system, conditioned by the available data. EnKF is of course one such method. However, other ensemble methods have been defined and studied.

It has been mentioned (equations [1.4]–[1.5]) that, in the linear and Gaussian case, an ensemble of independent realizations of the conditional probability distribution $P(x|z) = \mathcal{N}(x^a, \mathbf{P}^a)$ can easily be obtained. Perturb the data z following the probability distribution of the data error, and perform the analysis (equation [1.2]) on the perturbed data, for instance, by minimizing an objective function of form [1.14].

This approach can of course be implemented with nonlinear model and observation operators, even if there is no guarantee as to the Bayesianity of the ensembles that will be obtained. It has been implemented under the name of

Ensemble of Data Assimilations (EDA) at ECMWF and Météo-France. It has not been implemented there for the full assimilation, but only using a simplified model, in order to obtain a sample of analyzed states that are used at the end of the assimilation window, together with a static estimate of the background error covariance matrix, for defining the matrix $\mathbf{P}_0^b$ to be used in the objective function $\mathscr{J}_S(\xi_0)$ (equation [1.43]). Concerning the Bayesianity of the ensembles produced by EDA in nonlinear situations, Jardak and Talagrand (2018) have performed experiments on nonlinear systems, in particular on the chaotic system defined by Lorenz (1996). They have found that, as long as can be judged from statistics performed on the obtained ensembles, EDA produces, for the same ensemble size, ensembles that are as good an estimate of the real state and of the associated uncertainty as in the linear and Gaussian case, in which EDA is exactly Bayesian.

A number of ensemble algorithms, which may borrow from both the variational and the Kalman filter approach, have been defined, many of which are described and discussed in Chapter 7 of Asch et al. (2016). Variational assimilation requires the use of the adjoint of the assimilating model. Development, validation and maintenance of an adjoint code, although it is a rather straightforward task, require time and resources. The possibility of avoiding that task, or at least of significantly reducing its cost, has led to the development of new algorithms. Liu et al. (2008, 2009), following ideas initially put forth by Robert et al. (2005), instead of performing the minimization of an objective function in the entire state space, performed it in a relatively small subspace. The idea is that, if the dimension of the minimizing space is small enough, the gradient of the objective function with respect to the control variable can be computed by finite perturbation of the latter, thus avoiding the need for the adjoint of the dynamical model (or of possibly nonlinear observation operators). The gradient can then be computed through the transpose of a small dimension explicitly computed Jacobian matrix. This approach, of which numerous variants exist, is commonly called *4DEnVar*. Liu et al. (2008, 2009) minimized a quadratic approximation of the objective function, which leads to small-dimension matrix computations. The question of the definition of the initial background error covariance matrix remains, and the new question of how to choose the small dimension minimization space arises. A number of possible solutions, including for instance running an EnKF in parallel with the variational assimilation, have been studied.

Along the same idea of restricting the minimization to an appropriately defined subspace, Bocquet and Sakov (2014) have introduced the *Iterative Ensemble Kalman Smoother* (IEnKS). The gradient of the objective function with respect to the control variable is again computed by finite difference in a small dimension control space, a finite difference approximation of the Hessian of the objective function being determined at the same time. Once the minimizing solution has been determined, the control subspace is regenerated, on the basis of a Gaussian hypothesis, from the approximate Hessian. The IEnKS has been shown, on nonlinear systems of relative small dimensions, to outperform, in terms of final accuracy, both Strong Constraint Variational Assimilation and the EnKF.

1.4.4. *Stability and instability*

Experience shows that various algorithms that have been used for ensemble assimilation produce useful results for relatively small ensemble sizes, of the order of a few tens or at most a few hundreds, while they are used with numerical models that have state dimensions on the order of 10^{3-5} or more. This is fortunate in that it is largely what makes them useful, but it is also somewhat surprising that these relatively small dimensions can lead to useful results. Also, it raises the question of what a good choice of ensemble size is.

Consider for simplicity the case of a perfect model. Over the assimilation window, the error along the unstable components of the flow will increase, while the error along the stable modes will decrease. As a result, at the end of a variational assimilation (assuming that the observations are uniformly distributed over time), the residual error in the minimizing solution will be concentrated in the stable modes at the beginning of the assimilation window, and in the unstable modes at the end of the window. This aspect was studied by Pires et al. (1996), who found in numerical experiments that this was indeed the case. More recently, Gurumoorthy et al. (2017) and Bocquet et al. (2017) have rigorously proven that, in the case of an exact linear model, the error in the Kalman Filter concentrates asymptotically in the unstable manifold of the model. A consequence is that, along an assimilation window, late observations bring information on the unstable modes, which have developed over the window, and are therefore easier to observe. Early observations, on the other hand, bring information on the stable modes of the flow. This can be particularly important for a posteriori assimilation. Observations performed at the beginning and at the end of the assimilation window do not contain the same type of information and cannot be considered as interchangeable. Having observations distributed in time is not the same as having the same quantity of observations, with the same accuracy, concentrated at one time. The former says more on the stable or unstable character of the flow and of the components of the flow along the corresponding modes.

It results that an important aspect of assimilation is to monitor in one way or another and to prevent as much as can be done, the development of instabilities in the course of assimilation. Also, in the case of variational assimilation as well as of any form of smoother, it is important to monitor and control the stable components of the flow. Trevisan et al. (2010) have shown that concentrating the control variable of Variational Assimilation in the unstable subspace of the flow (identified by an ensemble integration of the model itself, which concentrates the dispersion of the forecasts in that subspace) can be more efficient than keeping the control variable in the entire state space.

Of course, in a linear case, Kalman Filter in its standard form, if implemented correctly, will explicitly describe the evolution of the uncertainty through equation [1.32]. In a nonlinear case, if a realistic evolution of the uncertainty is included in the assimilation (for instance, in an Ensemble Filter), the growth of instability should be monitored and taken into account in the analysis step of the filter.

However, the important fact is that in geophysical flows, the number of unstable components is relatively small. In large-scale meteorology and oceanography, the dominant instabilities result from the latitudinal temperature gradient. At any time, there are only a small number (typically a few tens) of "centers of action" where instabilities develop. Moreover, the physical process of instability, the so-called *baroclinic instability*, is fundamentally the same in those various centers of action. This is basically the reason that makes it possible to implement ensemble assimilation, in either meteorological or oceanographical applications, with relatively small ensembles.

All simplified forms of Kalman Filter or Variational Assimilation that have been presented above are based on the restriction of the estimation to a small dimension subspace of the entire state space. No systematic study of how that subspace must be chosen (in particular, in terms of the current situation of the flow) seems to have been performed. However, it is clear from the preceding discussion that it must include at least the directions along which instabilities have developed in the recent past.

1.5. Particle filters

All the algorithms that have been described up to now are fundamentally based on the linear and Gaussian approach described in section 1.2. This approach has been heuristically (and successfully) extended to moderately nonlinear and non-Gaussian situations. However, the basic equations that are used in those algorithms are fundamentally of the form [1.9]–[1.10] or [1.14], and all algorithms that have been used up to now in large-scale meteorological or oceanographical applications are of the same general form.

Particle filters are totally independent of any linear or Gaussian hypothesis (see, for example, Doucet et al. (2001) for an introduction). As their name suggests, they are ensemble methods in which the state of the system is described at any time, and as in EnKF, by a set of points in state space. However, a probability is now attached to each of these points, and it is both the particles and their associated probabilities that evolve in the course of the assimilation.

We give here a simple description of particle filters, without going into full mathematical details. Let us assume that a background ensemble $\{x_l\}$ $(l = 1, \ldots, L)$ is available at an observation time (we drop again the time index), together with associated probabilities, which we denote $\{p_l^b \equiv P(x_l)\}$ $(l = 1, \ldots, L)$. Let y be the observation vector at the same time. In order to update the probabilities with y and to obtain new probabilities p_l^a, we use Bayes' law, viz.,

$$p_l^a \equiv P(x_l|y) \propto P(y|x_l)P(x_l) \quad [1.50]$$

where the proportionality coefficient, if required, can be obtained from the condition that $\sum_l p_l^a = 1$. The conditional probability $P(y|x_l)$ (the so-called *likelihood*) can easily be obtained if, for instance, y is of form [1.18], the probability distribution of the error ϵ being independent of the probability distribution of the x_l's. Then $P(y|x_l) = P[\epsilon = y - H(x_l)]$. It is seen that, in this analysis step, the particles x_l are not modified, but their probabilities are.

In the forecast phase to the next observation time, the particles x_l are evolved according to the model equation [1.42] (or through any process that is consistent with the probability distribution defined by the weights p_l^a; see, for example, Carrassi et al. (2018)). If a random model error is present, an independent realization of that error has to be included in each of the L model integrations. This is easy to achieve if the model error is supposed to be independent of the distribution of the x_l's. In the forecast phase, the particles x_l are modified, but their probabilities are not.

It is clear that this procedure is independent of any linear or Gaussian hypothesis. As described, it depends, as do the standard Kalman Filter and most of the variants of the latter, on a hypothesis of independence in time of the errors (in precise mathematical terms, the appropriate condition is a condition of *Markovianity*). It is clearly Bayesian, at least in the sense that, if it is initialized from an ensemble of independent realizations of an initial probability distribution, it will produce at any later stage a sample of independent realizations of the probability distribution of the system state at that later stage, conditioned by the information (observations and model) that has been used in the meantime.

Particle filters are very efficient in small dimension systems, but when the dimension increases, it is observed that the ensembles tend to collapse, all the probability tending to concentrate on one particle. This is a manifestation of the *curse of dimensionality*, which is the fact that it is in practice very costly, if not simply impossible, to describe probability distributions in large dimensional spaces. Snyder et al. (2008) estimate that, in order to avoid ensemble collapse, the size of the ensemble must grow exponentially with the dimension of the state space. Active research is being pursued for Particle Filter schemes that might circumvent the curse of dimensionality. A general approach, called *regularization*, is to remove from the ensemble particles with low probability, and to replace them by appropriately chosen new particles in the region of phase space with high probability. For a review of particle filters as applied to high-dimensional systems, see, for example, van Leeuwen (2017).

1.6. Artificial intelligence

The continuing growth of the quantity of available data and of computing power has stimulated in many fields the development of AI and ML. These are methods that can produce knowledge through pure statistical processing of large amounts of data, without use of (or in addition to) a priori explicitly stated physical laws (or, for that matter, biological or economical laws, or any explicit statement relative to the behavior of the system of interest). ML is largely based on the use of *neural networks*, which establish explicit numerical links between given sets of "inputs" and "outputs". Atmospheric science requires in all its aspects large amounts of data and powerful computing, and is therefore a potential field of applications of these new developments. Data assimilation and ML, even if they are at present distinctly different in the way they are implemented and in the algorithms they use, are basically two approaches for achieving the same purpose, viz., extract as useful information as possible from large sets of data. In the context of atmospheric science, and in relation with data assimilation, ML is being considered for two classes of applications (which are not mutually exclusive): (1) identification of laws governing the evolution of the observed system (see, for example, Brajard et al. (2020)); and (2) identification of observation operators. Observation operators have not been discussed much above, but they relate the observations to the state variables of the assimilating model (equation [1.18]). They are often meant to represent complex, poorly known, physical processes, sometimes through their statistical impact at a spatial or temporal scale that is larger than the scale at which the observations are performed. An example is the interaction between clouds and radiation, which is poorly known even at the microscale at which it takes place, and has to be represented in assimilation at a much larger scale. We can hope that the new possibilities offered by AI and ML will lead to significant progress in, at least, the identification of observation operators. For more on the use of ML in the context of data assimilation, see, for example, Geer (2021).

1.7. Extensions and applications

When numerical weather prediction was first developed, in the early 1950s, what mattered was of course the forecast itself. The definition of the initial conditions, performed originally as a simple interpolation from observing stations to model gridpoints, was a very minor task. But then difficulties gradually arose, which led to including more and more components in the process of definition of the initial conditions: need for ensuring realistic spatial scales, together with approximate geostrophic balance, in the initial fields, introduction of observations that are distributed in time, with the need for using the forecast model even before the start of the forecast (it is on that occasion that the word *assimilation* was coined), introduction of observations that are linked to the model through complex, sometimes poorly known, relationships (for instance satellite observations), and need for monitoring and controlling the instabilities in the system. This resulted in major technical and numerical developments, and assimilation has now become a very substantial part of the whole process of numerical weather prediction. One day of assimilation now requires the same amount of computing resources as for five days of prediction or more (including ensemble prediction). There is little doubt that the gradual development of more and more powerful methods for assimilation has significantly contributed, together with the improvement of numerical models and of the observing system, to the improvement of the quality of the meteorological forecasts.

However, even in the world of meteorological prediction, assimilation has extended much beyond the simple definition of the initial conditions of a forecast. It is a basic diagnostic in the assessment of the quality of a model. The degree to which a model is capable to fit the observations through assimilation is one measure, among others, of the quality of the model. Removing a particular set of observations from an assimilation experiment gives a direct measure of the information content of that set of observations. Assimilation is systematically used for *Observing Systems Simulation Experiments* (OSSEs), intended at evaluating a priori the gain that possible future observing systems could bring on the knowledge of the state of the atmosphere and on the quality of the ensuing forecasts. Assimilation is also at the core of *reanalyses*, in which past observations are assimilated with present models and algorithms (Kalnay et al. 1996). Such reanalysis programs now process observations going back as far as the beginning of the 20th century (Laloyaux et al. 2018). Reanalyses are very useful, even fundamental, for climatic studies, especially now in the context of an evolving climate. However, they are also used for making reforecasts on past situations, which give additional and interesting insight on the performance and quality of present models.

Beyond atmospheric science, assimilation has naturally extended to many other aspects of geophysical fluid dynamics. Oceanography has already been mentioned. Not only do there exist programs for assimilation of observations of the ocean, but interactive assimilation of both meteorological and oceanographic observations in coupled models is of great potential interest, for instance, for seasonal prediction, and is the subject of active research. The difficulty here is the gap between typical time scales of the atmospheric and oceanic dynamics, which does not allow direct use of algorithms such as the ones that are described above. A discussion of the associated problems can be found in Penny and Hamill (2017).

Assimilation has now been used in almost all branches of climate science. We can mention atmospheric chemistry, hydrology, paleoclimatology, glaciology, the dynamics of the biosphere and the vegetation cover. It has become extraterrestrial with the study of the dynamics of planetary atmospheres. Outside strictly climatic science, it has been used to the study of seismology, plate tectonics and magnetism (not only terrestrial, but also solar), volcanology and to applications which actually go beyond physical and chemical sciences. Chapters 11 and 12 of Asch et al. (2016) give examples of applications of assimilation of observations in physical and chemical sciences, as well as in human and social sciences. Also, for a recent example, we can mention a work by Evensen et al. (2021) on the dynamics of the COVID-19 epidemic.

1.8. References

Asch, M., Bocquet, M., Nodet, M. (2016). *Data Assimilation: Methods, Algorithms, and Applications*. SIAM, Philadelphia, PA.

Bocquet, M. (2011). Ensemble Kalman filtering without the intrinsic need for inflation. *Nonlinear Proc. Geoph.*, 18, 735–750.

Bocquet, M. and Sakov, P. (2014). An iterative ensemble Kalman smoother. *Quarterly Journal of the Royal Meteorological Society*, 140, 1521–1535.

Bocquet, M., Raanes, P.N., Hannart, A. (2015). Expanding the validity of the ensemble Kalman filter without the intrinsic need for inflation. *Nonlinear Proc. Geoph.*, 22, 645–662.

Bocquet, M., Gurumoorthy, K.S., Apte, A., Carrassi, A., Grudzien, C., Jones, C.K. (2017). Degenerate Kalman filter error covariances and their convergence onto the unstable subspace. *SIAM/ASA Journal on Uncertainty Quantification*, 5, 304–333.

Bouttier, F. (1994). Sur la prévision de la qualité des prévisions météorologiques. PhD Thesis, Université Paul-Sabatier, Toulouse.

Brajard, J., Carassi, A., Bocquet, M., Bertino, L. (2020). Combining data assimilation and machine learning to emulate a dynamical model from sparse and noisy observations: A case study with the Lorenz 96 model. *Journal of Computational Science*, 44, 101171.

Buehner, M., Houtekamer, P.L., Charette, C., Mitchell, H.L., He, B. (2010a). Intercomparison of variational data assimilation and the ensemble Kalman filter for global deterministic NWP. Part II: One-month experiments with real observations. *Monthly Weather Review*, 138, 1567–1586.

Buehner, M., Houtekamer, P.L., Charette, C., Mitchell, H.L., He, B. (2010b). Intercomparison of variational data assimilation and the ensemble Kalman filter for global deterministic NWP. Part I: Description and single-observation experiments. *Monthly Weather Review*, 138, 1550–1566.

Carrassi, A., Bocquet, M., Bertino, L., Evensen, G. (2018). Data assimilation in the geosciences: An overview of methods, issues, and perspectives. *Climate Change*, 9(5), e535.

Cosme, E. (2015). Méthodes d'assimilation rétrospective de données et application à l'océanographie. HDR Thesis, Université Grenoble Alpes, Grenoble.

Courtier, P., Thépaut, J.-N., Hollingsworth, A. (1994). A strategy for operational implementation of 4D-Var, using an incremental approach. *Q. J. R. Meteorol. Soc.*, 120, 1367–1387.

Daley, R. (1991). *Atmospheric Data Analysis*. Cambridge University Press, Cambridge.

Derber, J. and Bouttier, F. (1999). A reformulation of the background error covariance in the ECMWF global data assimilation system. *Tellus*, 51A, 195–222.

Doucet, A., de Freitas, N., Gordon, N. (eds) (2001). An introduction to sequential Monte Carlo methods. In *Sequential Monte Carlo Methods in Practice*. Springer, Berlin.

Evensen, G. (1994). Sequential data assimilation with a non-linear quasi-geostrophic model using Monte Carlo methods to forecast error statistics. *J. Geophys. Res.*, 99(C5), 10143–10162.

Evensen, G. (2009). *The Ensemble Kalman Filter*. Springer, Berlin.

Evensen, G., Amezcua, J., Bocquet, M., Carrassi, A., Farchi, A., Fowler, A., Houtekamer, P.L., Jones, C.K., de Moraes, R.J., Pulido, M. et al. (2021). An international initiative of predicting the SARS-CoV-2 pandemic using ensemble data assimilation. *Foundations of Data Science*, 3(3), 413–477.

Gaspari, G. and Cohn, S.E. (1999). Construction of correlation functions in two and three dimensions. *Q. J. Roy. Meteor. Soc.*, 125(554), 723–757.

Geer, A. (2021). Learning earth system models from observations: Machine learning or data assimilation? *Phil. Trans. R. Soc. A.*, 379(2194). doi: 10.1098/rsta.2020.0089.

Ghil, M. and Malanotte-Rizzoli, P. (1991). Data assimilation in meteorology and oceanography. *Adv. in Geophys.*, 33, 141–266.

Ghil, M., Cohn, S., Tavantzis, J., Bube, K., Isaacson, E. (1981). Applications of estimation theory to numerical weather prediction. In *Dynamic Meteorology-Data Assimilation Methods*, Bengtsson, L., Ghil, M., Källén, E. (eds). Springer, Berlin.

Gurumoorthy, K.S., Grudzien, C., Apte, A., Carrassi, A., Jones, C.K. (2017). Rank deficiency of Kalman error covariance matrices in linear time-varying system with deterministic evolution. *SIAM J. Control Optim.*, 55, 741–759.

Houtekamer, P. and Mitchell, H.L. (2001). A sequential ensemble Kalman filter for atmospheric data assimilation. *Mon. Weather Rev.*, 129, 123–137.

Jardak, M. and Talagrand, O. (2018). Ensemble variational assimilation as a probabilistic estimator – Part 2: The fully non-linear case. *Nonlin. Processes Geophys.*, 25, 589–604. doi: 10.5194/npg-25-589-2018.

Jones, R.H. (1965). Optimal estimation of initial conditions for numerical prediction. *J. Atmos. Sci.*, 22, 658–663.

Kalman, R. (1960). A new approach to linear filtering and prediction problems. *J. Fluid. Eng.*, 82, 35–45.

Kalnay, E. (2002). *Atmospheric Modeling and Data Assimilation and Predictability*. Cambridge University Press, Cambridge.

Kalnay, E., Kanamitsu, M., Kistler, R., Collins, W., Deaven, D., Gandin, L., Iredell, M., Saha, S., White, G., Woollen, J. et al. (1996). The NCEP/NCAR 40-year reanalysis project. *Bull. Amer. Meteor. Soc.*, 77, 437–471.

Klinker, E., Rabier, F., Kelly, G., Mahfouf, J.-F. (2000). The ECMWF operational implementation of four-dimensional variational assimilation. III: Experimental results and diagnostics with operational configuration. *Q. J. R. Meteorol. Soc.*, 126, 1191–1215.

Laloyaux, P., de Boisseson, E., Balmaseda, M., Bidlot, J.-R., Broennimann, S., Buizza, R., Dalhgren, P., Dee, D., Haimberger, L., Hersbach, H. et al. (2018). Cera-20c: A coupled reanalysis of the twentieth century. *Journal of Advances in Modeling Earth Systems*, 10(1172), 97–110. doi: 10.1029/2018MS001273.

Le Dimet, F.X. and Talagrand, O. (1986). Variational algorithms for analysis and assimilation of meteorological observations: Theoretical aspects. *Tellus*, 38A, 97–110.

Le Gland, F., Monbet, V., Tran, V.-D. (2011). Large sample asymptotics for the ensemble Kalman filter. In *The Oxford Handbook of Nonlinear Filtering*, Crisan, D. and Rozovskii, B. (eds). Oxford University Press, Oxford.

van Leeuwen, P. (2017). Particle filters for nonlinear data assimilation in high-dimensional systems. *Annales de la faculté des sciences de Toulouse : Mathématiques*, 6(26), 1051–1085. doi: 10.1007/BF02427924.

Lewis, J. and Derber, J.C. (1985). The use of adjoint equations to solve a variational adjustment problem with advective constraints. *Tellus*, 37A, 309–322.

Liu, C., Xiao, Q., Wang, B. (2008). An ensemble-based four-dimensional variational data assimilation scheme. Part I: Technical formulation and preliminary test. *Mon. Weather Rev.* 136, 3363–3373.

Liu, C., Xiao, Q., Wang, B. (2009). An ensemble-based four-dimensional variational data assimilation scheme. Part II: Observing system simulation experiments with advanced research WRF (ARW). *Mon. Weather Rev.*, 137, 1687–1704.

Lorenz, E.N. (1996). Predictability: A problem partly solved. In *Seminar on Predictability*. ECMWF Reading, Berkshire.

Meng, Z. and Zhang, F. (2008). Tests of an ensemble Kalman filter for mesoscale and regional-scale data assimilation. Part IV: Comparison with 3DVAR in a month-long experiment. *Mon. Weather Rev.*, 136, 3671–3682.

Penenko, V. and Obraztsov, N.N. (1976). A variational initialization method for the fields of the meteorological elements. *Soviet Meteorol. Hydrol.*, 11, 1–11.

Penny, S. and Hamill, T.M. (2017). Coupled data assimilation for integrated earth system analysis and prediction. *Bull. Amer. Meteor. Soc.*, 98, ES169–E172.

Petersen, D. (1968). On the concept and implementation of sequential analysis for linear random fields. *Tellus* 20, 673–686.

Pham, D. (2001). Stochastic methods for sequential data assimilation in strongly nonlinear systems. *Mon. Weather Rev.*, 129, 1194–1207 [Online]. Available at: https://journals.ametsoc.org/view/journals/mwre/129/5/1520-0493_2001_129_1194_smfsda_2.0.co_2.xml.

Pham, D., Verron, J., Roubaud, M.C. (1998). Singular evolutive extended Kalman filter for data assimilation in oceanography. *J. Marine Systems*, 16, 323–340. doi: 10.1016/S0924-7963(97)00109-7.

Pires, C., Vautard, R., Talagrand, O. (1996). On extending the limits of variational assimilation in nonlinear chaotic systems. *Tellus*, 48A, 96–121.

Rabier, F., Järvinen, H., Klinker, E., Mahfouf, J.-F., Simmons, A. (2000). The ECMWF operational implementation of four-dimensional variational assimilation. I: Experimental results with simplified physics. *Q. J. R. Meteorol. Soc.*, 126, 1143–1170.

Robert, C., Durbiano, S., Blayo, E., Verron, J., Blum, J., Le Dimet, F.X. (2005). A reduced-order strategy for 4D-VAR data assimilation. *J. Marine Systems*, 57, 70–82.

Snyder, C., Bengtsson, T., Bickel, P., Anderson, J.L. (2008). Obstacles to high-dimensional particle filtering. *Mon. Weather Rev.*, 136, 4629–4640.

Talagrand, O. (1997). Assimilation of observations, an introduction. *J. Meteor. Soc. Japan*, 75(1B), 191–209.

Talagrand, O. and Miyakoda, K. (1971). The assimilation of past data in dynamical analysis. Part II. *Tellus*, 23, 318–327.

Thépaut, J., Hoffman, R.N., Courtier, P. (1993). Interactions of dynamics and observations in a 4-dimensional variational assimilation. *Mon. Weather Rev.*, 121, 3393–3414.

Trevisan, A., D'Isidoro, M., Talagrand, O. (2010). Four-dimensional variational assimilation in the unstable subspace and the optimal subspace dimension. *Q. J. R. Meteorol. Soc.*, 136, 487–496.

Verlaan, M. and Heemink, A. (1997). Tidal flow forecasting using reduced rank square root filters. *Stochastic Hydrol. Hydraul.*, 11, 349–368. doi: 10.1007/BF02427924.

Verlaan, M. and Heemink, A. (2001). Nonlinearity in data assimilation applications: A practical method for analysis. *Monthly Weather Review*, 129, 1578–1589.

2

Ensemble Data Assimilation in Volcanology

Mary Grace BATO[1], Virginie PINEL[2] and Yajing YAN[3]

[1] *JPL, Caltech, Pasadena, USA*

[2] *ISTerre, IRD, CNRS, Université Grenoble Alpes, Université Savoie Mont Blanc, Le Bourget-du-Lac, France*

[3] *LISTIC, Université Savoie Mont-Blanc, Annecy, France*

2.1. Volcano monitoring and eruption forecasting

Accurate forecasting of volcanic eruption is one of the primary goals of volcanology. What is the chance of a certain volcano erupting today? How big will the eruption be? Indeed, eruption forecasting is considered as one of the most challenging fields in volcanology, not only because of the complex, nonlinear behavior and intrinsic unpredictability of volcanoes (e.g. Sparks 2003), but also unlike in weather forecasting, we have a lack of direct observation on what is exactly happening underground. Although eruption forecasting is often only associated with the onset of the eruption, it is more than that. Eruption forecasting includes accurate and timely delivery of information about (1) if, when and where the eruption will occur; (2) what will be the size and volume of the eruption; (3) how long will it last; and (4) what possible impending catastrophic or transient events might take place. All these pieces of information are necessary to help the civil protection groups and decision-makers properly mitigate risks and assess hazards.

Current practices that lead to successful eruption forecasting are mostly based on empirical pattern recognition, which relies on combining monitoring data with information from global volcanic databases, understanding of a volcano's past behavior (i.e. geological and historical records) and scientific expertise based on experience and local knowledge about the volcano (Segall 2013). Except for a recent study (Rivalta et al. 2019), probabilistic maps of future eruptive vents are always derived empirically from the volcano's past behavior and current monitoring (e.g. (Marzocchi et al. 2008)). In general, we can divide eruption forecasting into two phases: (1) long-range and (2) short-range forecasting. Long-range forecasting involves interpretations of the evidence of magma emplacement at depth in storage zones (i.e. increasing pressure), which can take up to several years or decades (Phase I on Figure 2.1) and often requires a good monitoring network of instruments installed around an active volcano. Tracking subsurface pressure changes allows us to determine the moment of rupture of the chamber walls and can thus provide information about when the magma will further migrate after being stored at depth. Short-range forecasting is focused on tracking the migration of magma once the magma chamber ruptured. It could be within weeks, days or even minutes prior to the eruption (Phase II on Figure 2.1). Determining if, when and where the magma will reach the surface (e.g. Cannavò et al. 2015; Guldstrand et al. 2018), as well as characterizing the potential magnitude of the eruption, is the first crucial information needed to prepare and manage an impending disaster.

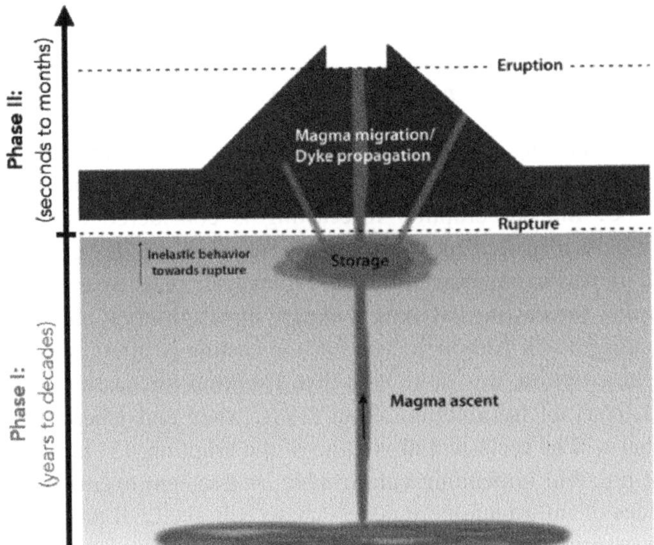

Figure 2.1. *Simplified schematic view of a magma plumbing system feeding volcanoes. For a color version of this figure, see www.iste.co.uk/yan/remotesensing.zip*

For many volcanoes, ground deformation is a powerful indicator of both their long-term and short-term behaviors. Deformation can be used to detect precursors earlier than what volcanic gas and seismic monitoring can. Dzurisin (2003) believes, including the authors of this work, that the next step is to formulate a deformation monitoring strategy, particularly at volcanoes where magma accumulates aseismically due to subsurface processes that are not detectable at the surface by other monitoring techniques. Ground deformation can be measured using various geodetic techniques. However, nowadays, satellite radar interferometry (i.e. Interferometric Synthetic Aperture Radar [InSAR]) and global position systems (i.e. GNSS) are the most common and easiest tools utilized to measure ground deformation on volcanoes (see section 2.2.3).

Traditionally, volcanologists interpret geodetic data using simple kinematic models to infer the properties of the magma reservoirs such as the source location, geometry and volume change (Anderson and Segall 2013). Volcano kinematic models provide a way to interpret ground surface displacements at a given epoch in terms of the "source strength" at depth, without considering the processes, forces and mechanisms at the origin of the displacement field. However, kinematic models often neglect important information (e.g. pressure evolution inside the magma chamber) that can give rise to the temporal behavior of the dataset. Recently, major advancement in eruption forecasting involves the development of more sophisticated and realistic process-based models that are coupled with diverse datasets to relate properties of the volcanic systems (Jaupart and Allègre 1991; Segall 2013). More than a decade ago, Sparks (2003) emphasized the need to evolve from forecasting based on empirical pattern recognition to forecasting based on the understanding of the physics of causative processes and that the resulting forecast should be expressed in probabilistic terms to account for uncertainties. Recent studies in Mount St. Helens have successfully demonstrated such an idea by applying a probabilistic Bayesian-based inversion technique through the Markov chain Monte Carlo (MCMC) method into deterministic (process-based) models and geophysical observations such as lava dome extrusion volume, GNSS data and CO_2 emissions (Anderson and Segall 2013; Wong et al. 2017). However, there are certain caveats when implementing Bayesian-based inversion techniques, which may include (1) expensive computational cost; (2) inefficiency of incorporating immense amounts of data in real time; (3) model errors are often ignored; and (4) difficulty in estimating time-varying parameters.

In this chapter, we will attempt to address the challenges in implementing classic inversion methods by demonstrating the potential of data assimilation as a complementary tool in tracking the state of the volcanic system and in estimating uncertain model parameters using both synthetic and real datasets during inter-eruptive periods. Focusing on long-term forecasting (Phase I on Figure 2.1), we will begin with a section describing the strategy for tracking the magma pressure evolution in storage zones using a dynamic model for the interaction of two connected magma reservoirs and geodetic observations, which can be linked to the model. Then, we will discuss the potential of using data assimilation via the ensemble Kalman filter (EnKF) through synthetic experiments. Finally, we will present our proposed framework to assimilate actual ground deformation data measured at Grímsvötn volcano, Iceland and showcase the significant contribution of using EnKF in forecasting the volcanic system state and estimating a time-varying parameter.

2.2. Ensemble data assimilation

Sequential data assimilation is an inversion strategy that allows us to efficiently combine dynamic forward model and time series of observations. On the one hand, predictions given by the forward model incorporate uncertainties due to (1) the choices and/or limitations related to the model physics and parameters, including errors associated with the assumptions, theory and/or conceptualizations within the underlying equations; 2) the computational grid and its discretization; and (3) the time step or numerical methods used to solve the mathematical equations. On the other hand, uncertainties are also present in the observations due to the instrument itself, different perturbations during data acquisition and noise generated during the pre- and post-processing of the data. Moreover, in most cases, observations are not spatially or temporally complete because of the limitations in data acquisition and/or the data extraction methods used. Indeed, data assimilation takes advantage of the complementary information provided by the dynamic model and the observations while incorporating their respective uncertainties. It corrects model predictions whenever observations are available in order to provide a model state trajectory as accurate as possible.

Although data assimilation is already an operational technique in numerical weather predictions, its application to solve problems in volcanology is still very much in its infancy. Early efforts to follow the behaviors of volcanoes are focused on applying various Kalman filter (KF) spin-offs (e.g. Fournier et al. 2009; Shirzaei and Walter 2010; Gregg and Pettijohn 2016; Zhan and Gregg 2017; Zhan et al. 2017; Albright et al. 2019; Xue et al. 2020) using kinematic models (e.g. Mogi 1958; McTigue 1987; Yang et al. 1988) and geodetic data. It is only until recently that dynamic models have been proposed to assess the potential of volcanic eruptions and to understand subsurface volcanic processes by assimilating deformation data (e.g. Bato et al. 2017, 2018).

In the following sections, we will detail our sequential data assimilation strategy (section 2.2.1) and the necessary ingredients to perform this: the dynamic model (section 2.2.2) and the deformation observations (section 2.2.3).

2.2.1. *Volcanic data assimilation using the ensemble Kalman filter*

One of the most popular methods in data assimilation is the EnKF. It is an ensemble-based data assimilation technique developed by Evensen (1994, 2003) as an alternative to solve the limitations of the classic Kalman Filter. The main characteristic lies on the use of N-ensemble of realizations to construct a Monte Carlo approximation of the mean and covariance of the state vector (i.e. vector containing all the model parameters to be improved by data assimilation). In this study, we design a simple and first-order framework based on the works of Evensen (1994, 2003) and Burgers et al. (1998). We track and forecast the state of the volcanic system in quasi-real-time during inter-eruptive periods and estimate uncertain parameters of the dynamic model.

In general, sequential data assimilation consists of two steps: (1) the forecast step; and (2) the update step (also known as the analysis step). EnKF begins with the ensemble generation. N-realizations of uncertain model parameters are generated according to an a priori distribution. Then, Monte Carlo simulations are performed while running forward the dynamic model for each ensemble member, resulting in the ensemble of model state forecast X, that is, $X = \begin{bmatrix} x_{1,1} & x_{1,2} & x_{1,N_n} \\ x_{2,1} & \ddots & \vdots \\ \cdots & & \\ x_{N_x,1} & \cdots & x_{N_x,N_n} \end{bmatrix}$, where N_x is the number of state variables and N_n is the ensemble size.

The forecast step is carried out by model integration under the control of the model operator representing the physical processes governing the system. The model predicts the state at the current instance from the model state at the previous step, as expressed in equation [2.1].

$$X^f_{t_{i+1}} = \mathcal{M}X^a_{t_i} + q_{t_i} \tag{2.1}$$

f and a denote the forecast and analysis, respectively, $\mathcal{M}$ is the model operator derived from equations [2.7] and [2.8] that relates the state of the system at time, t_i to t_{i+1} and q is the model error. At each time step, the error covariance, P^f, of the model forecast (i.e. an $N_x \times N_x$ matrix) can be obtained using:

$$P^f = \overline{(X^f - \overline{X^f})(X^f - \overline{X^f})^T} \tag{2.2}$$

The update step consists of the correction of the model prediction by the observations to obtain a more accurate estimation of the model state. The observations, D, that is, $D = \left[d_1, d_2, ..., d_{N_m}\right]^T$, where N_m is the total number of observations, can be anything related to the true model state, $x^\dagger$, through the observation operator $\mathcal{H}$.

$$D = \mathcal{H}x^\dagger + \epsilon \tag{2.3}$$

Note that the observation error, ϵ, allows the derivation of the error covariance of the observations, R, that is, $R = \mathbb{E}(\epsilon\epsilon^T)$.

Updating X^f is straightforward and can be performed by first computing the Kalman gain, K, using equation [2.4]. The Kalman gain represents the magnitude at which the incoming observation corrects the model forecast in equation [2.1].

$$K = P^f\mathcal{H}^T(\mathcal{H}P^f\mathcal{H}^T + R)^{-1} \tag{2.4}$$

To formalize the update, we use the following equation:

$$X^a = X^f + K(D - \mathcal{H}X^f) \tag{2.5}$$

The result of the update is called the analysis, X^a. It is the linear sum of the model forecast and the correction introduced by the observation, given the uncertainties of the observation and the model prediction. In general, the analysis must be at least statistically as accurate as any of the individual observations or the model prediction.

The error covariance of the analysis, P^a, is given by equation [2.6].

$$P^a = \overline{(X^a - \overline{X^a})(X^a - \overline{X^a})^T} \tag{2.6}$$

2.2.2. *The dynamic model*

Various time-dependent models have been suggested to capture the dynamic behavior of connected storage zones and the transport of magma toward the surface (e.g. Mériaux and Jaupart 1995; Lengliné et al. 2008; Michaut 2011; Le Mével et al. 2016). We chose the most generic and simplest model, the two-magma reservoir model (Figure 2.2) by Reverso et al. (2014), based on the recent evidences and interpretations of magmatic storage zones beneath volcanoes. Indeed, most volcanoes are thought to be fed by multiple magmatic storage zones located at various depths that are stacked vertically beneath the edifice (e.g. Elsworth et al. 2008; Bagnardi and Amelung 2012). The Reverso et al. (2014) model consists of two reservoirs, connected by a hydraulic pipe, in an elastic half-space. The deeper reservoir is fed by a constant magma inflow, which corresponds to the bottom boundary condition of the system. For simplicity, the magma is assumed to be incompressible but compressibility effects could be introduced easily in the shallower reservoir to account for degassing (Rivalta and Segall 2008). The model is characterized by a set of geometrical and rheological parameters (Table 2.1) and solves for the temporal evolution of the magma overpressures, ΔP_s and ΔP_d, within the shallow and the deep reservoirs, respectively (i.e. the overpressure is defined as the excess pressure within a magma reservoir as compared to a lithostatic state of reference). As shown by Reverso et al. (2014), this simple model provides a consistent explanation for the temporal behavior of the inter-eruptive displacement measured at Grímsvötn volcano, Iceland after its last three eruptions (1998, 2004 and 2011). The initial and transient exponential behavior indicates the refilling of the shallow reservoir by the deeper one after the eruption. Once the system has readjusted, a constant displacement rate can be observed due to the steady magma inflow at the bottom of the system.

The two-magma reservoir model has overpressure values within the shallow and deep reservoirs at a given time, t_{i+1}, that can be derived from their values at the previous time step, t_i, using the following discrete time step equations (Reverso et al. 2014; Bato et al. 2017):

$$\frac{\Delta P_{s_{t_{i+1}}} - \Delta P_{s_{t_i}}}{t_{i+1} - t_i} = \frac{G{a_c}^4}{8\mu\gamma_s H_c {a_s}^3}((\rho_r - \rho_m)gH_c + \Delta P_{d_{t_i}} - \Delta P_{s_{t_i}}) \quad [2.7]$$

$$\frac{\Delta P_{d_{t_{i+1}}} - \Delta P_{d_{t_i}}}{t_{i+1} - t_i} = \frac{G}{\gamma_d \pi {a_d}^3} Q_{in} - \frac{\gamma_s {a_s}^3}{\gamma_d {a_d}^3} \frac{\Delta P_{s_{t_{i+1}}} - \Delta P_{s_{t_i}}}{t_{i+1} - t_i} \quad [2.8]$$

γ_s and γ_d indicate the shapes (i.e. spherical or sill-like) of the shallow and deep reservoirs, respectively. The discrete formula stays valid as long as the time interval $\Delta t = t_{i+1} - t_i$ remains small compared to the time constant τ of the system given by $\tau = \frac{8\mu H_c {a_s}^3 {a_d}^3 \gamma_s \gamma_d}{G{a_c}^4({a_s}^3\gamma_s + {a_d}^3\gamma_d)}$.

The radial, u_R, and vertical, u_z, displacements observed at the surface can be expressed as follows:

$$u_R(r, t_i) = \frac{(1-v)}{G} r\left(\alpha_s \frac{{a_s}^3}{R_s^3} \Delta P s_{t_i} + \alpha_d \frac{{a_d}^3}{R_d^3} \Delta P_{d_{t_i}}\right) + u_R(r, t_0) \quad [2.9]$$

$$u_z(r, t_i) = \frac{(1-v)}{G} \left(H_s \alpha_s \frac{{a_s}^3}{R_s^3} \Delta P s_{t_i} + H_d \alpha_d \frac{{a_d}^3}{R_d^3} \Delta P_{d_{t_i}}\right) + u_z(r, t_0) \quad [2.10]$$

where the shapes (i.e. spherical or sill-like) of the shallow and deep reservoirs are represented by two geometrical constants, α_s and α_d, respectively. Note that if we assume that magma reservoir overpressures are initially set at zero ($\Delta P_{s_0} = 0$ and $\Delta P_{d_0} = 0$), then the initial displacement are null ($u_R(r, t_0) = 0$ and $u_z(r, t_0) = 0$).

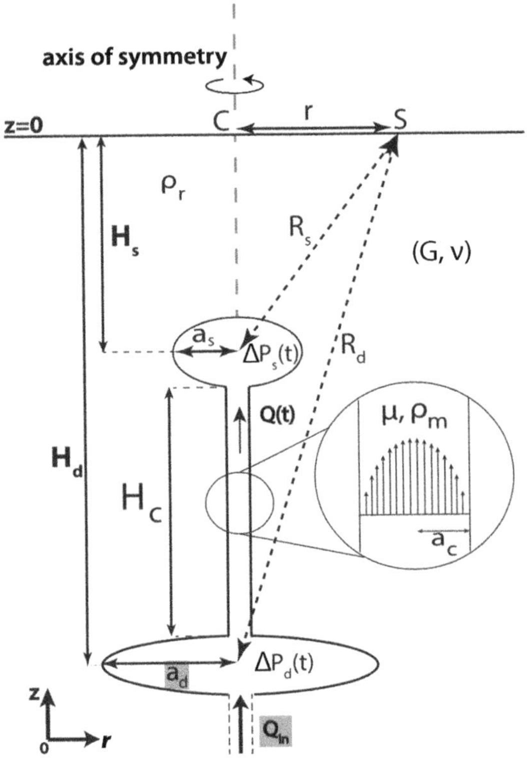

Figure 2.2. *Schematic diagram of the two-magma reservoir model, modified after Reverso et al. (2014). The magma inflow rate at the bottom chamber, Q_{in}, and the radius of the deep reservoir, a_d, are the two parameters considered as uncertain in this study. Observations (vertical and horizontal displacements) are recorded at the surface at a given location, S, characterized by its distance, r, from the center of the volcanic system, C. $R_s = \sqrt{r^2 + H_s{}^2}$ and $R_d = \sqrt{r^2 + H_d{}^2}$ are distances between S and the shallow and deep reservoirs, respectively*

Provided that we have access to the deformation fields related to the activity of the volcano inferred from geodetic data (see section 2.2.3), then equations [2.9] and [2.10] create a link between the dynamic model and the observations, necessary for data assimilation.

Parameters	Description	Value	Unit
Geometry			
a_c	Radius of the conduit.	1.6	m
a_d	Radius of the deep reservoir.	2.2	km
a_s	Radius of the shallow reservoir.	2.0	km
H_d	Depth of the deep reservoir.	35.0	km
H_s	Depth of the shallow reservoir.	3.0	km
γ_d	1.0 for Mogi point source, $8(1-\nu)/3\pi$ for sill.	1.0	
γ_s	1.0 for Mogi point source, $8(1-\nu)/3\pi$ for sill.	$8(1-\nu)/3\pi$	
α_d	1.0 for Mogi point source, $(4{H_d}^2)/(\pi {R_d}^2)$ for sill.	1.0	
α_s	1.0 for Mogi point source, $(4{H_s}^2)/(\pi {R_s}^2)$ for sill.	$(4{H_s}^2)/(\pi {R_s}^2)$	
Physics			
G	Shear modulus.	81.9	GPa
ν	Poisson's ratio.	0.25	
μ	Viscosity.	2,000.0	Pa s
ρ	Density contrast, $(\rho_r - \rho_m)$.	300.0	kg m^{-3}
g	Gravity.	9.81	m s^{-2}
Basal condition			
Q_{in}	Magma flow rate in the deep reservoir.	0.02	km^3 y^{-1}
Initial conditions			
ΔP_{s0}	Value of the shallow reservoir overpressure at t_0.	0.0	Pa
ΔP_{d0}	Value of the deep reservoir overpressure at t_0.	0.0	Pa

Table 2.1. *Model parameters and true values assigned for the synthetic simulations. The highlighted parameters are the ones considered as uncertain in this study. For a color version of this table, see www.iste.co.uk/yan/remotesensing.zip*

2.2.3. *Data observations*

2.2.3.1. *Global navigation satellite system*

Global navigation satellite system (GNSS) is a generic term for satellite navigation systems including, but not limited to, GPS (United States), GLONASS (Russia), Galileo (Europe) and Beidou (China). It is a constellation of satellites, where each satellite transmits radio navigation signals about its location and time at current intervals. The information is then intercepted by a receiver, which computes the distance between each satellite based on the timing of the arrival of information. The position of the receiver can be derived when these distances are combined with ephemerides for the satellite orbits and satellite clock corrections (i.e. available from commercial vendors or IGS). Multi-GNSS receivers are becoming common, allowing simultaneous measurements from multiple navigation satellite systems, therefore providing high precision measurements. Typically, the positions are averaged in either static (i.e. daily or hourly) or kinematic mode (i.e. series of independent position at each epoch). Static mode is preferred for permanent GNSS measurements, whereas kinematic mode is favored for the campaign type of measurement (i.e. field surveying). An accuracy of $\sim$ 1 mm for the horizontal component and $\sim$ 3 mm for the vertical component is typical from static daily post-processed data (Freymueller et al. 2015). Note that the vertical component is always less accurate than the horizontal component, because it is more sensitive to atmospheric delays, glacial isostatic adjustments as well as to tidal changes. Many active volcanoes around the world have at least one continuous GNSS installed to monitor its activity in real time. Indeed, a network of permanent GNSS scattered around the volcano is an ideal monitoring system. Campaign GNSS measurements are also sometimes performed at least once a year to provide better spatial coverage; however, performing regular field measurements are logistically expensive, especially when the volcano of interest is located in a remote area.

2.2.3.2. *Radar interferometry*

InSAR circumvents the spatial coverage constraint of GNSS. SAR technology uses the microwave region of the electromagnetic spectrum to image the Earth, thus it can work even with the presence of clouds. SAR is usually mounted on a satellite or aircraft while orbiting the Earth. Satellite-based SAR systems, for example, move in a sun-synchronous polar orbit. SAR satellites operate as an active system (i.e. no need for the radiation emitted by the sun), thus they have the capabilities to image the Earth during both day and night. Most of the operational SAR systems today are in L, C and X-bands (Pinel et al. 2014). SAR images are formed when microwave signals emitted by the satellite are backscattered by the ground after it has been illuminated by a radar pulse. Each pixel in a radar image is a complex number that contains the amplitude and phase information.

A SAR image alone cannot provide information about any surface change. At least two properly co-registered SAR images taken at two distinct dates are needed. We call the first image acquired by the satellite as the reference image, whereas the second acquisition as the secondary image. A pair of SAR images can be useful to detect any surface change by calculating the amplitude and/or phase variation.

The principle of InSAR exploits the phase differences between the reference and secondary images. If there is a local surface displacement that occurred between the two acquisitions, a concentric pattern of signals or "fringes" can be observed. The resulting phase map is basically the sum of the phases contributed by the topography (ϕ_{topo}), orbit of the satellite (ϕ_{orb}) (i.e. ϕ_{topo} and ϕ_{orb} are due to the small difference of satellite positions between the two acquisitions), ground displacement (ϕ_{disp}), atmospheric delay (ϕ_{atm}) and other noise (ϕ_{ϵ}) (Massonnet and Feigl 1998).

$$\Delta\phi = \phi_{topo} + \phi_{disp} + \phi_{orb} + \phi_{atm} + \phi_{\epsilon} \quad [2.11]$$

In monitoring volcanic unrest, only the phase due to the ground displacement (ϕ_{disp}) is important, whereas the others are corrected and/or reduced in as much as possible (e.g. ϕ_{orb} can be removed by using the satellite's orbit data). In other cases where mapping and possibly quantifying the volume of erupted volcanic deposits (e.g. lava flow emplacement and lava dome formation) are of priorities, the phase due to topography is rather retained (Bato et al. 2016), provided that the temporal baseline (i.e. difference between the acquisition dates) remains small such that the magnitude of the displacement can be ignored.

The ground displacement retrieved from an interferogram is in fact the displacement along the line-of-sight (LOS) of the satellite within the time covered by the two SAR acquisitions. In order to derive the three-component displacement vector, a more discriminating dataset characterized by several interferograms with different acquisition geometries (i.e. different look angles and orbit directions) are required (e.g. Wright et al. 2004; Yan et al. 2012). Although in practice, having only a pair of ascending and descending images is sufficient under some assumptions (i.e. NS component is neglected, since current satellite systems are least sensitive to the displacement in this direction) to retrieve the horizontal and vertical components of the displacement.

Nowadays, there are several algorithms available to produce a time series of ground deformation using InSAR. Most of these multi-temporal InSAR processing algorithms are based on (1) Persistent Scatterer Interferometric Synthetic Aperture Radar (PSInSAR) (e.g. Perissin and Ferretti 2007; Hooper et al. 2004); (2) small baseline subsets (SBAS) (e.g. Berardino et al. 2002; Doin et al. 2011; Agram et al. 2013; Zhang et al. 2019); and (3) phase linking or phase triangulation (e.g. Ferretti et al. 2011; Samiei-Esfahany and Hanssen 2013; Fornaro et al. 2015). In Figure 2.3, we show an example of the cumulative and time series of ground deformation recorded at Sierra Negra volcano, Ecuador, derived from InSAR and GNSS data.

Indeed, InSAR is a valuable tool in deformation monitoring. Although it cannot beat the (near) real-time data capability of continuous GNSS, the temporal and spatial resolution of InSAR have greatly improved over the past years. The launch of the Sentinel-1A satellite in 2014 and Sentinel-1B in 2016 became the game changer for everyone working on deformation monitoring. Anyone can have access to free SAR data every 6 or 12 days, which can be crucial, especially during an on-going volcanic crisis when in situ instruments have stopped operating. This has been exemplified recently during the Taal 2020 eruption in the Philippines (Bato et al. 2021).

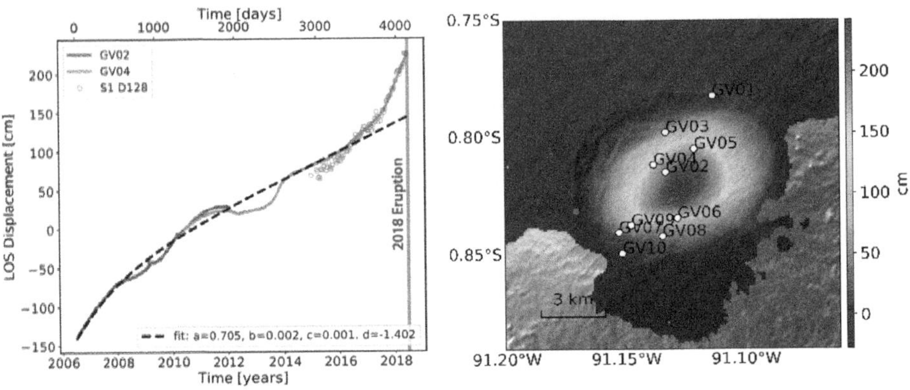

Figure 2.3. *Ground deformation measured at Sierra Negra volcano, Ecuador. (Left) GNSS and InSAR displacement time series along the radar line-of-sight (LOS). Solid lines are the GV02 (blue) and GV04 (orange) GNSS measurements projected in LOS that are only shown here for comparison with the InSAR-derived time series. Note that each GNSS measurement provides data in three dimensions (i.e. along the East, North and Vertical directions). The green circles represent the InSAR-derived time series at the location of GV04 station, calculated from Sentinel-1 descending track 128 using the Jet Propulsion Laboratory's ARIA processing system and MintPy (Zhang et al. 2019). The black broken line is the curve fit for the GNSS measurements using the generic form of the two-chamber model (i.e.* $A(1-e^{-Bt})+Ct+D$*). The red vertical line marks the June 26, 2018 eruption of Sierra Negra. (Right) InSAR cumulative displacement map from December 13, 2014 to June 19, 2018. White circles are the locations of the GNSS stations that monitor the ground deformation at the Sierra Negra volcano. For a color version of this figure, see www.iste.co.uk/yan/remotesensing.zip*

2.3. Potentiality assessment of volcanic data assimilation for eruption forecasting based on synthetic simulations

The main goal of this section is to evaluate the capability of EnKF to quantify the magma pressure evolution at depth from synthetic surface displacement data and thus to forecast pressure-based volcanic eruptions. Most of the results presented here are derived from a previous study by Bato et al. (2017) where they had performed both state estimation (deriving only the state variables ΔP_s and ΔP_d) and state-parameter estimation experiments (deriving the state variables: ΔP_s and ΔP_d as well as two uncertain parameters: a_d, the radius of the deep reservoir and Q_{in}, the basal magma inflow rate). In this chapter, we will only present the results of the state-parameter estimation, which are more useful and efficient in real cases.

2.3.1. *EnKF formulation*

The state vector can be expressed as:

$$X = \left[\Delta P_s, \Delta P_d, a_d, Q_{in}\right]^T \qquad [2.12]$$

where, again, ΔP_s is the overpressure in the shallow reservoir, ΔP_d is the overpressure in the deep reservoir, a_d is the radius of the deep reservoir and Q_{in} is the basal magma inflow rate. We focused on these particular uncertain model parameters characterizing the deeper part of the magma plumbing system, because they are the most difficult to recover using ground deformation data. We rewrite the dynamic model into matrix form, that is:

$$\begin{bmatrix} \Delta P_{s_{t_{i+1}}} \\ \Delta P_{d_{t_{i+1}}} \\ a_{d_{t_{i+1}}} \\ Q_{in_{t_{i+1}}} \end{bmatrix} = \begin{bmatrix} 1 - C_1\Delta t & C_1\Delta t & 0\,0 \\ C_1C_2\Delta t & 1 - C_1C_2\Delta t & 0\,0 \\ 0 & 0 & 1\,0 \\ 0 & 0 & 0\,1 \end{bmatrix} \begin{bmatrix} \Delta P_{s_{t_i}} \\ \Delta P_{d_{t_i}} \\ a_{d_{t_i}} \\ Q_{in_{t_i}} \end{bmatrix} + \begin{bmatrix} C_1A_1\Delta t \\ (A_2 - C_1C_2A_1)\Delta t \\ 0 \\ 0 \end{bmatrix} \qquad [2.13]$$

such that the model operator is now $\mathcal{M} = \begin{bmatrix} 1 - C_1\Delta t & C_1\Delta t & 0\,0 \\ C_1C_2\Delta t & 1 - C_1C_2\Delta t & 0\,0 \\ 0 & 0 & 1\,0 \\ 0 & 0 & 0\,1 \end{bmatrix}$.

The relationships between the surface displacements and the overpressures are described by equations [2.9] and [2.10], if we let $\Gamma = \frac{1-\nu}{G}$, $D_s = \alpha_s \frac{a_s^3}{R_s^3}$ and $D_d = \alpha_d \frac{a_d{}^3}{R_d^3}$, then we can rewrite them into a matrix form:

$$\begin{bmatrix} u_{R_{t_i}} \\ u_{z_{t_i}} \end{bmatrix} = \begin{bmatrix} \Gamma D_s r & \Gamma D_d r & 0\,0 \\ \Gamma D_s H_s & \Gamma D_d H_d & 0\,0 \end{bmatrix} \begin{bmatrix} \Delta P_{s_{t_i}} \\ \Delta P_{d_{t_i}} \\ a_d \\ Q_{in} \end{bmatrix} \qquad [2.14]$$

The observation vector is given by $D = \begin{bmatrix} u_{R_{t_i}} \\ u_{z_{t_i}} \end{bmatrix}$ and the observation operator, $\mathcal{H} = \begin{bmatrix} \Gamma D_s r & \Gamma D_d r & 0\,0 \\ \Gamma D_s H_s & \Gamma D_d H_d & 0\,0 \end{bmatrix}$, where r are the locations of the observations (i.e.

distance between point location at S and the center of the volcanic system, C, (see Figure 2.2) and H_s and H_d are the depths of the shallow and deep reservoirs, respectively. The observation error covariance is $R = \begin{bmatrix} \sigma_{u_R}^2 & 0 \\ 0 & \sigma_{u_z}^2 \end{bmatrix}$, where σ_{u_R} and σ_{u_z} are the radial and vertical displacement errors.

Note that a_d is included in the observation operator, $\mathcal{H}$. This can add complexity to the assimilation process, since a_d is treated as an uncertain model parameter that may contribute as another source of error coming from the observation operator. To incorporate this error in the assimilation process, we also calculate for $\mathcal{H}$ at each time step, t_{i+1}, and for each ensemble member, n, by sampling from the distribution of a_d obtained at the previous analysis step such that:

$$D_{t_{i+1}} = \mathcal{H}_{t_{i+1},n}[X^f_{t_{i+1},n}] \qquad [2.15]$$

2.3.2. *Synthetic observations*

The observations are the vertical and the radial surface displacements observed at a given time, t_i, and at a given distance, r, from the axis of symmetry of the system. From a model simulation with a given set of parameters listed in Table 2.1, we generate a set of synthetic overpressures using the analytic solution in Reverso et al. (2014):

$$\Delta P^{\dagger}_{s_{t_i}} = A(1 - e^{-\frac{t_i}{\tau}}) + \frac{GQ_{in}}{\pi({a_s}^3\gamma_s + {a_d}^3\gamma_d)} t_i + \Delta P^{\dagger}_{s_{t_0}} \qquad [2.16]$$

$$\Delta P^{\dagger}_{d_{t_i}} = \frac{\gamma_s A}{\gamma_d}(1 - e^{-\frac{t_i}{\tau}}) + \frac{GQ_{in}}{\pi({a_s}^3\gamma_s + {a_d}^3\gamma_d)} t_i + \Delta P^{\dagger}_{d_{t_0}} \qquad [2.17]$$

with $A = \frac{{a_d}^3\gamma_d}{{a_s}^3\gamma_s + {a_d}^3\gamma_d}[\Delta P^{\dagger}_{d_{t_0}} - \Delta P^{\dagger}_{s_{t_0}} + (\rho_r - \rho_m)gH_c - \frac{8\gamma_s Q_{in}\mu H_c {a_s}^3}{\pi {a_c}^4({a_s}^3\gamma_s + {a_d}^3\gamma_d)}]$. These overpressures are considered as the true values and are then used as input in equations [2.9] and [2.10] in order to produce the synthetic displacements $d^{\dagger}_m$, that is, $d^{\dagger}_m = \left[u^{\dagger}_{R_m}, u^{\dagger}_{z_m}\right]^T$.

White Gaussian noise, that is, $\mathcal{N}(\mu_u = 0, \sigma_u^2)$, is then added to generate the synthetic displacements, where σ_u represents the observation error along the radial and vertical directions. We use the typical GNSS instrument errors, 1 mm and 10 mm, for u_R and u_z, respectively.

2.3.3. *Experiment setup*

The state vector is designed as shown in equation [2.12]. The assimilation time interval, Δt, is fixed to 2 days, which is much smaller than the true time constant τ (i.e. $\tau \simeq 0.11$ y). The assimilation begins just after an eruption when the reservoirs are refilling and terminates after 500 time steps (i.e. $t_f \sim 2.74$ years). The initial values of magma overpressures are set to zero (see Table 2.1), assuming that both reservoirs had been fully depressurized by the previous eruption. The frequency of available observation, f_{obs}, is usually fixed to one; meaning that at each assimilation step (i.e. every 2 days), there is always two available observations. In the case wherein no observation ($f_{obs} = 0$) is available, the model prediction cannot be corrected by the observation. This particular case is hereafter called as the *free run* and is presented for comparison.

We used the stochastic EnKF approach where the observations are perturbed prior to assimilation (Burgers et al. 1998). Below we describe in detail the practical implementation of EnKF as summarized in Figure 2.4:

1) We start by defining an initial ensemble of 1000 members. For the ensemble generation, two distributions are considered for the uncertain parameters: (a) using a truncated-Gaussian distribution, wherein the mean of each distribution is centered on the true value of the uncertain parameters (hereafter called *unbiased*); and (b) using a Gaussian-prior distribution that does not include the true value of the uncertain parameters, the mean of each distribution being far from their true values (hereafter called *biased*).

2) For each of the ensemble members:

a) We run the forward model step-by-step from the initial condition until the data assimilation step t_{i+1}. The state vector is built as described in equation [2.12].

Tip 1: An inflation factor $\rho_{infl} \in [0, 1)$ can be multiplied to the ensemble of state variables, that is, $X^f = (1 + \rho_{infl})X^f$, to prevent the ensemble from collapsing to a single value. For all the synthetic cases performed, we used $\rho_{infl} = 0.1$ at each assimilation step. Note that prior to the application of an inflation factor, the forecast state vector must be centered.

Tip 2: In the state-parameter estimation using EnKF, the parameters are only updated by the covariance between them and the state variables. When doing so, randomly perturb the uncertain parameters by adding noise, $\mathcal{N}(0, \alpha_p^2)$, where α_p is the additive inflation. This would help the filter to explore more possible values since their values do not change during the forecast step, that is, $p^f_{t_{i+1},n} = p^a_{t_i,n}$. In our synthetic cases, we used an additive inflation of $\alpha_{a_d} = 5$ and $\alpha_{Q_{in}} = 0.005$ to tune the uncertain parameters during the assimilation. These are derived from empirical observations after several adjustments.

b) The error covariance of the forecast, P^f, is computed using equation [2.2].

c) We perturb the field observation vector using:

$$\hat{d}_m = d_m + \eta \qquad [2.18a]$$

$$\hat{D} = [\hat{d}_1, \hat{d}_2, ..., \hat{d}_{N_m}] \qquad [2.18b]$$

where η is a random variable with distribution $\mathcal{N}(0, R)$.

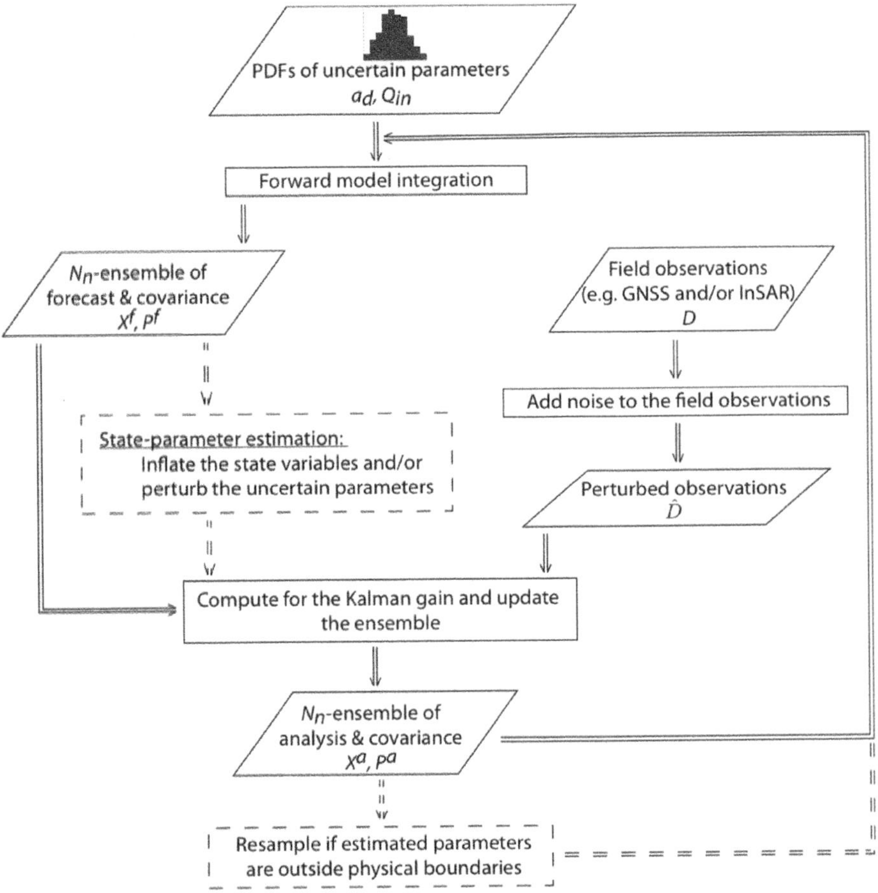

Figure 2.4. *The step-by-step EnKF strategy implemented in this study. The broken borders and lines imply that the step is optional*

d) The state vector is then updated using equation [2.5], replacing D with $\hat{D}$:

$$X^a = X^f + K(\hat{D} - HX^f) \qquad [2.19]$$

Tip 3: Resample the analysis state vector, X^a, using the strategy proposed by Evensen (2009) before moving to the next data assimilation time step if the updated values of the uncertain parameters fall beyond or exceed the boundary conditions.

e) The error covariance of the analysis, P^a, is obtained using equation [2.6].

2.3.4. *Results and discussions*

2.3.4.1. *State-parameter estimation*

The result of the state-parameter experiment shows that EnKF can track almost perfectly the true behavior of the overpressures as illustrated in Figure 2.5, where the prior distributions of a_d and Q_{in} are biased. In fact, the final values of the overpressures have as small as $0.0 - 0.01\%$ and $0.04 - 0.06\%$ error difference with respect to their true values for the shallow and deep overpressures, respectively.

In Figure 2.6, we plotted a_d and Q_{in} at each assimilation step using two different a priori distributions for the uncertain parameters. Note how the EnKF-estimated parameters converged well to their true values, regardless of how they were initialized. Bato et al. (2017) demonstrated that in the case of poorly initialized model parameters, the data fit worsens through time if state-only estimation is performed rather than state-parameter estimation. In Figure 2.7, we illustrate an improved prediction of the displacements via state-parameter estimation, thereby emphasizing the importance of including uncertain model parameters in the state vector during the assimilation.

2.3.4.2. *Joint assimilation of GNSS and InSAR-like datasets*

GNSS data have the advantage to capture the behavior of the overpressures because of its high temporal resolution as it is possible to obtain daily or sub-daily observations, whereas InSAR are less frequent but provide a better spatial information about the ground deformations and constraints on the uncertain parameters. In order to exploit the advantages of both datasets, we jointly assimilated the GNSS-like and InSAR-like data. Figure 2.8 shows how the evolution of the overpressures is well captured and how the uncertain model parameters converged to their true values. In fact for this case, in as early as ~ 6 months of jointly assimilating GNSS and InSAR, it may already be possible to forecast the long-term overpressure values that can later suggest whether a critical overpressure, an indicator whether or not an eruption may occur and is unique for each volcano, will most likely be achieved.

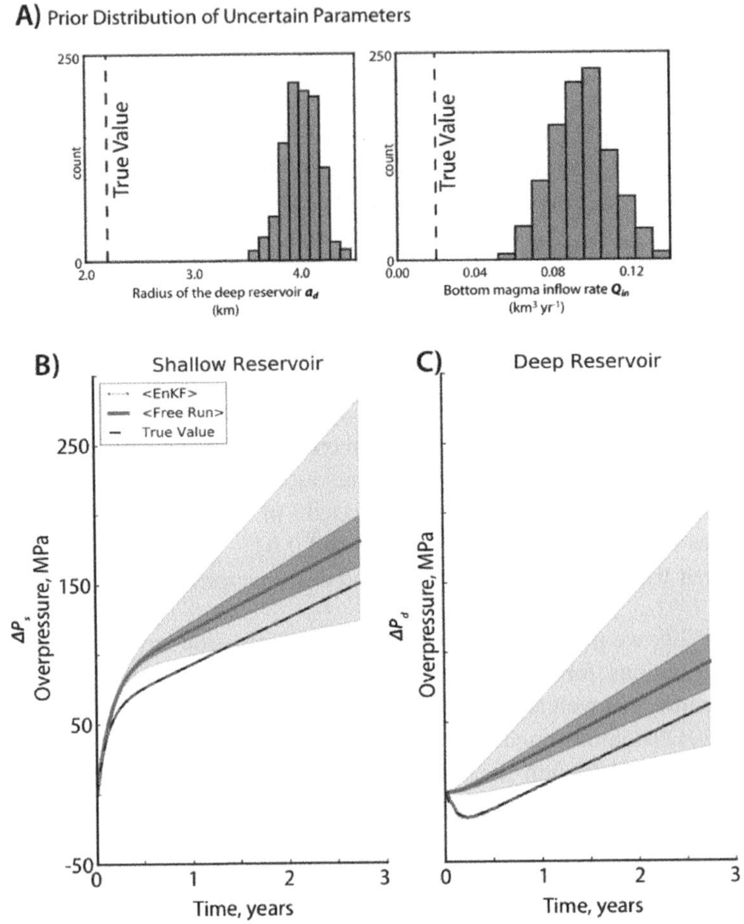

Figure 2.5. *The evolution of the overpressures after performing the state-parameter estimation given that the prior distributions of the uncertain parameters are Gaussian but are not centered on their true values (black broken lines). In (B) and (C), the black broken lines represent the true value of the overpressures; the green curves represent the result of the free run where the green solid line is the mean overpressure, the dark green fill is the spread (1σ) and the light green fill shows the full extent of the ensemble (i.e. [min, max]); the red curves represent the result of EnKF where the red solid line is the mean, the dark red fill is the spread (1σ) and the light red is the full extent of the ensemble. Note that the spread of the ensemble is very narrow for the assimilation case. For a color version of this figure, see www.iste.co.uk/yan/remotesensing.zip*

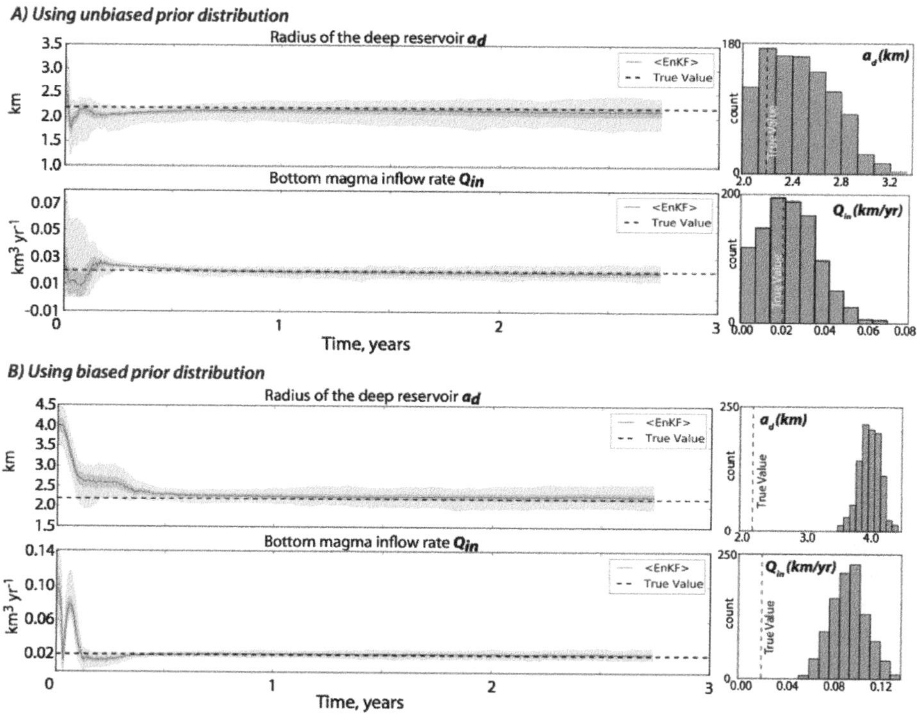

Figure 2.6. *The EnKF-predicted uncertain parameters after performing the state-parameter estimation, given that the prior distributions of the uncertain parameters are (A) unbiased or (B) biased. The solid red line is the mean value of the uncertain parameters. The dark-shaded and light-shaded red colors represent the spread (1σ) and the [min,max] of the ensemble, respectively. The true values are the black broken lines. For a color version of this figure, see www.iste.co.uk/yan/remotesensing.zip*

2.3.4.3. *Comparing with inversion*

Gregg and Pettijohn (2016) have previously compared data assimilation with inversion using two different models: a viscoelastic assumption for EnKF and an elastic one for the inversion. In our case, to be fair and consistent, the same forward dynamic model (the two-magma-chamber model) is used to compare the two techniques. We also use the same prior distributions for the two uncertain parameters considered: a_d and Q_{in} (Figure 2.9). We built a prior distribution that agrees with the prerequisites of both assimilation (optimized for Gaussian distribution) and inversion (working only when the true values of the uncertain parameters are included inside the range of the initial parameter distribution).

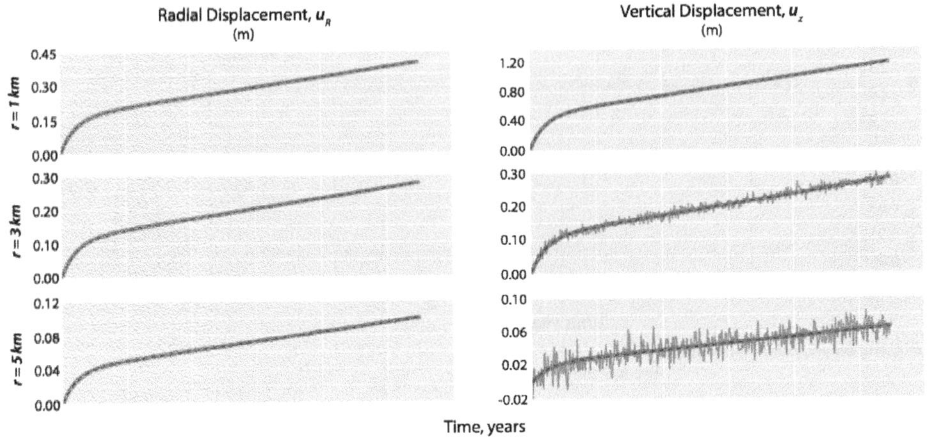

Figure 2.7. *Six of the 80 EnKF-estimated displacements after the state-parameter estimation (i.e. blue solid line), given that the prior distributions of the parameters are biased (Figure 2.6(B)). The synthetic observations used as D during assimilation are the noisy red lines that are more evident in the vertical component at far-field distances. The black broken lines correspond to the true values of the displacements. For a color version of this figure, see www.iste.co.uk/yan/remotesensing.zip*

For the EnKF, we performed the state-parameter estimation strategy as described in section 2.3.1, whereas we implemented a Bayesian-based estimation, that is, MCMC, for the inversion. MCMC is most useful in cases where models are nonlinear and expressing an analytical solution is not possible (Segall 2013). Note that we used 80 synthetic observations (i.e. 40 radial and 40 vertical displacements) that are uniformly (every 100 m) located at a distance $r = 1 - 4.9$ km with frequency of available observation consistent with the time interval (i.e. $\Delta t = 2$ days; $f_{obs} = 1$) for the two techniques.

In performing MCMC, the posterior distribution of the uncertain parameters is sampled given their a priori distribution using the forward model and a proposed likelihood distribution. Models are selected based on the Metropolis-Hastings rules, which always accept models that fit better to observations than the previous iteration and randomly accept those that do not fit to avoid being trapped to a local minima (Segall 2013). For example, at t_i, a set of a_d and Q_{in} values are drawn from their initial distributions, generating model predictions using the two-chamber model. These model predictions are then compared to the GNSS and/or InSAR data and are always accepted if the fit is better than the previous sampling iteration, creating the so-called Markov chain. The sampling iteration can be performed a thousand to a million times in order to build a full posterior distribution.

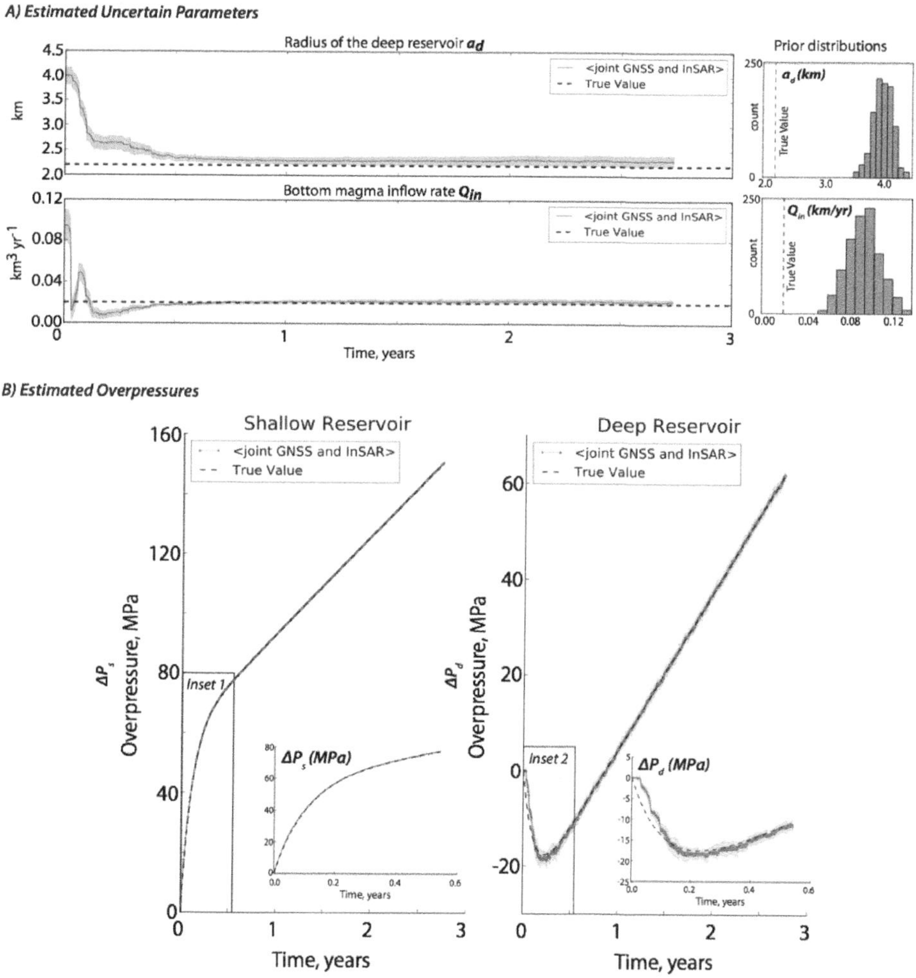

Figure 2.8. *The estimated (A) uncertain parameters and (B) overpressures after jointly assimilating GNSS (i.e. 10 observations that are assimilated every time step, $f_{obs} = 1$, with distance to the volcanic center ranging from 1 to 5 km) and InSAR-like observations (242 observations that are only assimilated every 12 days, $f_{obs} = 1/6$, with distance to the volcanic center ranging from 1 to $5\sqrt{2}$ km). The initial conditions of the uncertain parameters are the prior distributions in (A). The insets provide a magnified view of the overpressures at the start of the assimilation. The pink shade corresponds to the spread (1σ). The black broken lines represent the true values. For a color version of this figure, see www.iste.co.uk/yan/remotesensing.zip*

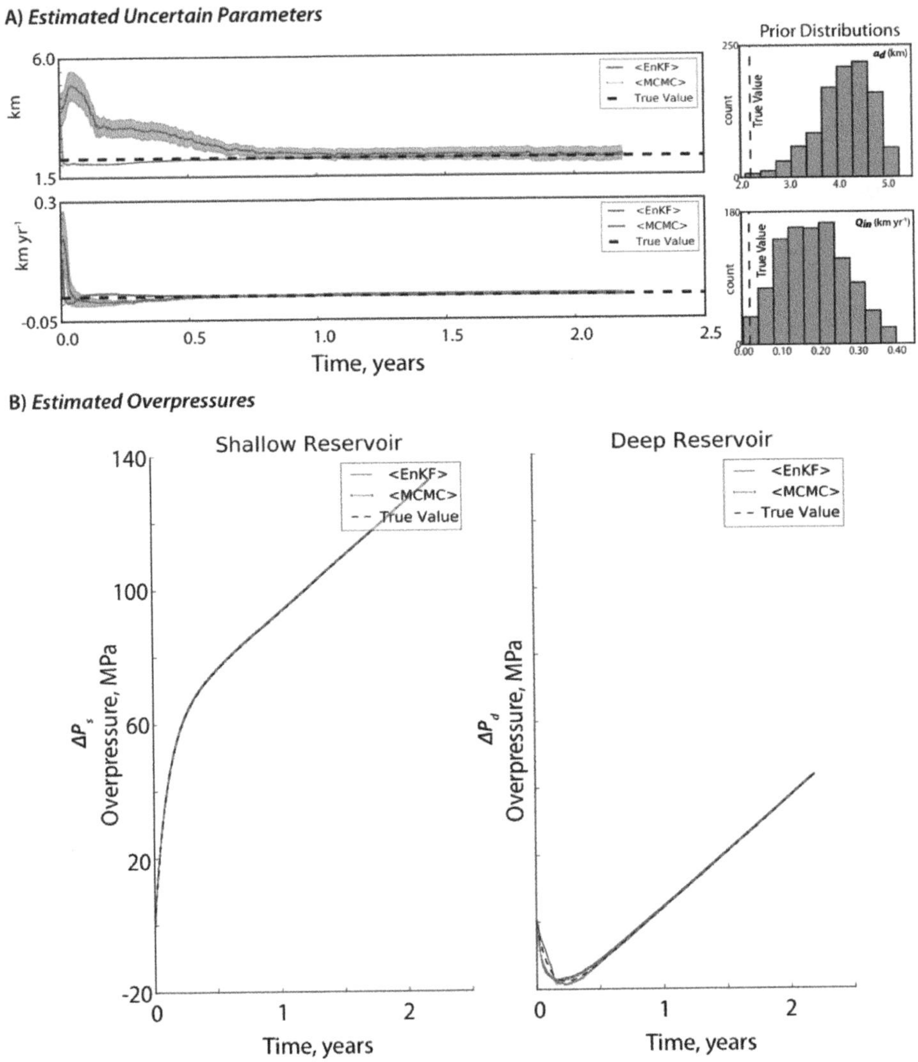

Figure 2.9. *Comparison between the EnKF-predicted and MCMC-predicted (A) uncertain model parameters and (B) overpressures. The PDFs in (A) are the prior distributions of a_d and Qin used during the data assimilation and Bayesian inversion. For the two techniques, we utilized 80 synthetic observations (i.e. combined radial and vertical displacements) located uniformly, every 10 m, at $r = 1–4.9$ km and are available every time step. The light blue and pink shades represent the spread (1σ) of the estimation. The true values are the black broken lines. For a color version of this figure, see www.iste.co.uk/yan/remotesensing.zip*

In the present case, we performed 11,000 MCMC sampling iterations at each time step and burned-in the first 10,000 so that in the end we have a similar ensemble size (e.g. 1,000) to that of the EnKF. Note that unlike in EnKF, which is sequential and only uses incoming observations to capture the temporal evolution of the overpressures, MCMC utilizes all the observations from t_0 up to the preferred time of observation t_i, thus becomes computationally expensive through time. We performed up to 400 time steps with interval similar to that in EnKF (i.e. $\Delta t = 2$ days).

Figure 2.9 presents the resulting overpressures and the parameter estimates after performing EnKF and MCMC. The shallow overpressure is well estimated by the two techniques, whereas the deep overpressure in both cases started from a deviated value but eventually converged to the true value. The MCMC estimates are also found smoother than the EnKF predictions, especially in the deep reservoir. Since EnKF always assumes that the dynamic model is uncertain and needs to be corrected by incoming observations, it then tends to closely follow the behavior of the observations, hence we observe a noisier estimation with EnKF. Note that in the MCMC inversion, epistemic model errors or uncertainties related to processes not included in the physical forward model, are not accounted (Segall 2013).

If we observe how the uncertain parameters are estimated by the two techniques, we will find that both are able to accurately estimate the true values by the end of the assimilation experiment, but MCMC converged faster. Take note, however, that the uncertain parameters we used in this study are static parameters. In the case of an evolving model parameter, which could be the case for the basal magma inflow (i.e. (Poland et al. 2012)), the inversion may not be the optimal method to use for estimation. We will elaborate this in section 4.4.

2.3.5. *Implications to real-time volcano monitoring*

While parameter estimation allows us to gain more knowledge about the plumbing system and the behavior of the volcano, in real-time crisis, one of the key variables to infer an impending effusive eruption is the overpressure. The EnKF strategy presented here uses a simple dynamic model that can be easily integrated with large amounts of real-time geodetic data, allowing us to quickly and accurately track the overpressures both in short-term and long-term periods. Assuming a statistic distribution for the threshold magma overpressure leading to reservoir wall rupture, based on the previous eruption for instance, the updated overpressure forecast provided by EnKF can be used to estimate the timing of an impeding eruption. Although the critical overpressure value is not always known, especially for volcanoes that do not erupt frequently, it depends on the rock strength that can be estimated and on the local stress field strongly influenced by the edifice geometry (Pinel and Jaupart 2003).

Another important challenge of the data assimilation approach is the availability of frequent data. Although for GNSS we can obtain daily to sub-daily observations, InSAR data are less frequent and are still dependent on the quality of the interferograms produced. Also, in reality, the 3D-displacement field vector from InSAR are not always retrievable due to the satellite's acquisition geometry. Furthermore, while we only used deformation data alone, the observation vector can include gas emission and seismic data for a more deterministic approach in forecasting (i.e. seismic data in particular can be used to estimate the timing of an eruption), as long as they can be related to the dynamic model used.

Interestingly, when it comes to near-real-time monitoring, it may be possible to use inversion and data assimilation jointly in order to accommodate vast amount of incoming data. MCMC allows faster prediction of non-evolving model parameters, whereas EnKF via state-estimation is easy to implement and only requires incoming observations. We will demonstrate this in the following section.

2.4. Application: The 2004–2014 inter-eruptive activity at Grímsvötn volcano, Iceland

We developed a two-step strategy using the radial surface displacement at Grímsvötn volcano to track the pressure evolution beneath it and to estimate a parameter that may suddenly change in time (i.e. basal magma inflow, Q_{in}). Step 1: we applied a Bayesian-based inversion with the MCMC algorithm to first constrain the non-evolving uncertain model parameters (e.g. a_d, H_d, C, $\delta\rho$, $\delta P_{d_{t_0}}$) and obtained a prior distribution for Q_{in} using only an initial subset of our inter-eruptive dataset (i.e. dotted green boxes in Figure 2.10(B). Step 2: we implemented EnKF following the strategy described in Figure 2.4 using the entire time series. Note that we separately performed this two-step method for each of the inter-eruptive cycle (i.e. 2004–2011 and 2011 up to the August 2016 rifting event).

After step 1, we only took the best-fit values of the non-evolving uncertain model parameters (e.g. a_d, H_d, C, $\delta\rho$, $\delta P_{d_{t_0}}$) from their resulting posterior distributions as constant inputs to step 2. For Q_{in}, as it can vary through time it remained as an uncertain parameter to estimate, so we redefined its distribution before performing EnKF (step 2) by taking its best-fit value as the mean and by setting a larger standard deviation (i.e. $\sim$ 0.003 km^3 year^{-1}). The result of forward modeling using these values is the free run and is represented by the green curves in Figure 2.10. The free run results in Figure 2.10 (bottom figure) shows a good data fit during the 2004–2011 inter-eruptive cycle but failed to fit the entire 2011 post-eruptive time series. We argue that this is because the magma inflow rate suddenly changed at some point in 2013. Thanks to data assimilation via EnKF, we were able to track this change as evidenced by the good fit to our data from 2004 to 2014 (i.e. red results in Figure 2.10, bottom plot). The assimilation of radial displacement using the strategy that we have developed not only results to a robust fit to the entire dataset but also enables us to closely follow the decreasing trend of Q_{in} in 2013. We obtain a minimum rate of 0.007 km^3 year^{-1} from the ensemble of Q_{in} estimates, which corresponds to a drop of 0.039 km^3 year^{-1} (85% decrease) relative to its prior value.

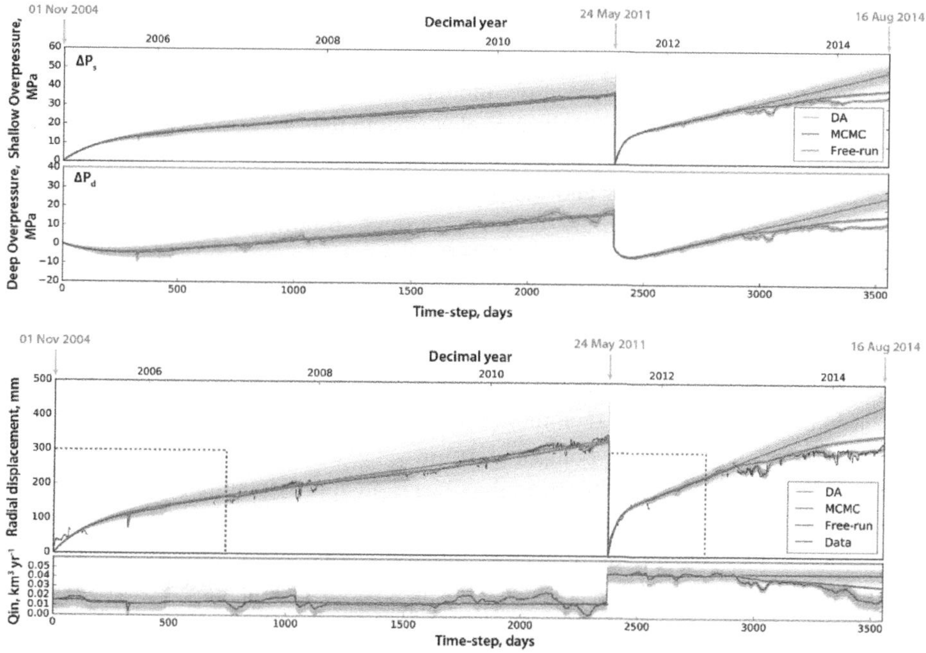

Figure 2.10. *(Top) The overpressure evolution inside the shallow and deep magma reservoirs beneath Grímsvötn volcano from 2004 to 2014. (Bottom) The displacement data fit and the magma inflow rate estimation from 2004 to 2014. The orange arrows mark the end of the eruptions in 2004 and 2011, as well as the start of the rifting event in August 2014. The black line is the actual radial displacement data. For both the top and bottom figures: the green curves are the result of the free-run after step 1, the red curves are the EnKF results after step 2 and the blue curves are the result of performing MCMC inversion, instead of EnKF in step 2. For a color version of this figure, see www.iste.co.uk/yan/remotesensing.zip*

Another interesting result is that if we follow the similar approach to track Q_{in} by first fixing the non-evolving uncertain parameters, but use an inversion approach (i.e. MCMC) as a second step instead of data assimilation, we find that MCMC was able to slightly detect the change in Q_{in} but the resulting fit with the data is not satisfactory (blue curves in Figure 2.10, bottom plot). The main difference comes from the fact that with MCMC, we invert at each given time step all the observations that are previously acquired considering an effective constant value of Q_{in} over the whole period (i.e. integral analytic formula), whereas with the assimilation strategy, we apply the differential equation between each time step such considering an evolving magma supply rate, Q_{in}.

In the classic inversion setup, the model used to interpret the data is assumed "perfect". In that case, the source of error is often only attributed to the data, either due to the perturbation in acquisition, instrument noise, data pre-processing or sum of these errors. However, in reality, models are embedded with noise and are oversimplified representations of the complex system that we observe. In section 2.3, we illustrated through a synthetic case that if the dynamic model used well explains the observed data time series and that there is no transient change in the uncertain model parameters (i.e. they are constant through time), both data assimilation through EnKF and inversion via MCMC can track the state variables (i.e. overpressures within the reservoirs) and also estimate the true values of the uncertain model parameters (i.e. basal magma inflow rate and radius of the deep reservoir).

However, if the uncertain model parameter varies through time, such as the case of Q_{in} in this study, EnKF is then more useful. In EnKF, the model error covariance, $P = \overline{(X - \overline{X})(X - \overline{X})^T}$, computed from a large number of perturbations of uncertain model parameters (for example, Q_{in}) is an approximation of the real model error. In practice, we use a large ensemble of models to represent the model error. Furthermore, by using a multiplicative inflation, which is a tuning step in data assimilation, underestimated model errors related to external process in EnKF and/or unaccounted source of model error are compensated.

2.4.1. *Implications of the change in magma supply rate at Grímsvötn*

The estimated decrease in the magma supply rate measured at Grímsvötn corresponds to a minimum deficit of 0.016 km^3 of magma for the Grímsvötn plumbing system, when during the same period, magma accumulation is expected at Bárdarbunga volcano (Gudmundsson et al. 2016). Bárdarbunga volcano is a neighboring volcano of Grímsvötn in Iceland that had an eruption in August 2014 (Sigmundsson et al. 2015; Gudmundsson et al. 2016). A similar anti-correlated behavior has previously been observed between Kilauea and Mauna Loa based on their eruptive activities, however characterized by isotopically different magma (Gonnermann et al. 2012). It was then interpreted as due to stress transfer over the 35 km distance separating the two systems through pore-pressure variations in a thin asthenospheric melt accumulation layer. Connections between volcanic systems have also been evidenced by geodetic observations previously but only for systems spaced less than 10 km apart (Biggs et al. 2016). Mechanisms invoked to explain nearby volcanic systems interactions include stress changes, lateral hydraulic connections and a common asthenospheric magma supply as for the Hawaiian case. In the case of Grímsvötn and Bárdarbunga, we can exclude the stress change effect due to the large distance separating the two systems (Biggs et al. 2016). We hypothesized two possible scenarios that caused the decrease in magma supply to Grímsvötn between October 2013 and August 2014: (1) the magma that feeds Grímsvötn's mid-crustal reservoir at $\sim$ 20 km depth was transported toward Bárdarbunga's volcanic system through an existing deep fracture; and (2) there was a drop in the relative pressure difference between Grímsvötn's mid-crustal reservoir and a much deeper reservoir shared by Bárdarbunga and Grímsvötn at more than 20 km depth. This sudden pressure drop is possibly due to the withdrawal of a large amount of magma in the shared reservoir of Grímsvötn and Bárdarbunga that may have been caused by the activation or reactivation of a connection between Bárdarbunga's shallow reservoir and its deep shared reservoir with Grímsvötn.

The possible magma inflow beneath Bárdarbunga, at least 10 months before the start of the dyke rifting event, was also evidenced by a slight increase in seismicity and deformation recorded at Bárdarbunga (Sigmundsson et al. 2020). However, given the large amount of magma mobilized by the Bárdarbunga eruption (i.e. 2 km^3), this magma inflow cannot be the only factor involved in the eruption initiation and buoyancy effects are probably involved (Sigmundsson et al. 2020). Subsequently, the withdrawal of magma and the gradual collapse of the caldera are both responsible to sustain the eruption for up to $\sim$ 6 months (Gudmundsson et al. 2016).

2.5. Conclusions and outlook

Will scientists be able, in the future, to provide a reliable eruption forecast similar to weather forecast, despite the lack of direct observation of the magmatic activities occurring below the Earth's surface?

Regardless of the important advancements in our observation systems and our understanding of volcanoes, there are many more processes occurring beneath the Earth that are yet to be unraveled. Volcanoes are complex systems and their physics are not fully understood (Poland and Anderson 2020). Slight changes in their system properties could actually result to sudden failure, hence, some systems may be inherently unpredictable (Sparks et al. 2012).

Indeed, the work that we have presented here is based on a simple conduit model and can be used for long-term forecasting. In the future, other assimilation strategies could be integrated into dynamic models of magma transport through magma intrusions or open conduit to address the issue of short-term forecasting. However, this is already a big leap toward a better understanding of volcanic behaviors. The remarkable potential of data assimilation that we have shown both in synthetic and real cases can open a whole new perspective in addressing current challenges in volcano eruption forecasting. Truly, if we want to provide an eruption forecast in the same way as we have a daily weather forecast, we need realistic and sophisticated process-based dynamic models that can be linked to diverse datasets such as gas emission and seismic data.

The advantage of using sequential data assimilation to follow the evolution of the overpressures and the deep magma supply rate is clear and represents a first successful application of EnKF in volcanology. Aside from the interest of forecasting volcanic eruptions, we have demonstrated that sequential data assimilation has also a real and unique potential to give insights into the deep plumbing system of volcanoes and the evolution of its bottom conditions through time.

2.6. Acknowledgments

We thank the ARIA team for activating and managing the ARIA Project Sentinel-1A/B SAR interferogram generation based on the European Commission Copernicus Sentinel-1 satellite SAR data products courtesy of the European Space Agency. Part of this research was carried out at the Jet Propulsion Laboratory, California Institute of Technology, USA, under a contract with the National Aeronautics and Space Administration (grant 281945.02.47.04.51). M.G. Bato receives support through the JPL R&TD program (project number 01STCR- R.18.245.034). V. Pinel acknowledges support through the MagmaPropagator project (contract ANR-18-CE92-0037-01).

2.7. References

Agram, P., Jolivet, R., Riel, B., Lin, Y., Simons, M., Hetland, E., Doin, M., Lasserre, C. (2013). New radar interferometric time series analysis toolbox released. *Eos, Transactions American Geophysical Union*, 94(7), 69–70.

Albright, J., Gregg, P., Lu, Z., Freymueller, J. (2019). Hindcasting magma reservoir stability preceding the 2008 eruption of Okmok, Alaska. *Geophysical Research Letters*, 46(15), 8801–8808.

Anderson, K. and Segall, P. (2013). Bayesian inversion of data from effusive volcanic eruptions using physics-based models: Application to Mount St. Helens 2004–2008. *Journal of Geophysical Research: Solid Earth*, 118(5), 2017–2037.

Bagnardi, M. and Amelung, F. (2012). Space-geodetic evidence for multiple magma reservoirs and subvolcanic lateral intrusions at Fernandina Volcano, Galapagos Islands. *Journal of Geophysical Research*, 117. doi: 10.1029/2012JB009465.

Bato, M.G., Froger, J., Harris, A., Villeneuve, N. (2016). Monitoring an effusive eruption at Piton de la Fournaise using radar and thermal infrared remote sensing data: Insights into the October 2010 eruption and its lava flows. *Geological Society, London, Special Publications*, 426(1), 533–552.

Bato, M.G., Pinel, V., Yan, Y. (2017). Assimilation of deformation data for eruption forecasting: Potentiality assessment based on synthetic cases. *Frontiers in Earth Science*, 5, 48.

Bato, M.G., Pinel, V., Yan, Y., Jouanne, F., Vandemeulebrouck, J. (2018). Possible deep connection between volcanic systems evidenced by sequential assimilation of geodetic data. *Scientific Reports*, 8(1), 1–13.

Bato, M.G., Lundgren, P., Pinel, V., Solidum, R., Daag, A., Cahulogan, M. (2021). The 2020 eruption and the large lateral dike emplacement at Taal volcano, Philippines: Insights from satellite radar data. *Geophysical Research Letters*, 48(7), e2021GL092803.

Berardino, P., Fornaro, G., Lanari, R., Sansosti, E. (2002). A new algorithm for surface deformation monitoring based on small baseline differential SAR interferograms. *IEEE Transactions on Geoscience and Remote Sensing*, 40(11), 2375–2383.

Biggs, J., Robertson, E., Cashman, K. (2016). The lateral extent of volcanic interactions during unrest and eruption. *Nature Geoscience*, 9(4), 308–311.

Burgers, G., Jan van Leeuwen, P., Evensen, G. (1998). Analysis scheme in the ensemble Kalman filter. *Monthly Weather Review*, 126(6), 1719–1724.

Cannavò, F., Camacho, A.G., González, P.J., Mattia, M., Puglisi, G., Fernández, J. (2015). Real time tracking of magmatic intrusions by means of ground deformation modeling during volcanic crises. *Scientific Reports*, 5, 10970.

Doin, M.-P., Guillaso, S., Jolivet, R., Lasserre, C., Lodge, F., Ducret, G., Grandin, R. (2011). Presentation of the small baseline NSBAS processing chain on a case example: The Etna deformation monitoring from 2003 to 2010 using Envisat data. In *Proceedings of the Fringe Symposium*, Fringe 2011 Workshop, Frascati.

Dzurisin, D. (2003). A comprehensive approach to monitoring volcano deformation as a window on the eruption cycle. *Rev. Geophys.*, 41(1), 1001. doi: 10.1029/2001RG000107.

Elsworth, D., Mattioli, G., Taron, J., Voight, B., Herd, R. (2008). Implications of magma transfer between multiple reservoirs on eruption cycling. *Geosciences School of Environmental Sciences*, 322(5899), 246–248.

Evensen, G. (1994). Sequential data assimilation with a nonlinear quasi-geostrophic model using Monte Carlo methods to forecast error statistics. *Journal of Geophysical Research: Oceans*, 99(C5), 10143–10162.

Evensen, G. (2003). The Ensemble Kalman Filter: Theoretical formulation and practical implementation. *Ocean Dyn.*, 53(4), 343–367.

Evensen, G. (2009). *Data Assimilation: The Ensemble Kalman Filter*. Springer Science & Business Media, New York.

Ferretti, A., Fumagalli, A., Novali, F., Prati, C., Rocca, F., Rucci, A. (2011). A new algorithm for processing interferometric data-stacks: SqueeSAR. *IEEE Transactions on Geoscience and Remote Sensing*, 49(9), 3460–3470.

Fornaro, G., Verde, S., Reale, D., Pauciullo, A. (2015). Caesar: An approach based on covariance matrix decomposition to improve multibaseline multitemporal interferometric SAR processing. *IEEE Transactions on Geoscience and Remote Sensing*, 53(4), 2050–2065.

Fournier, T., Freymueller, J., Cervelli, P. (2009). Tracking magma volume recovery at Okmok volcano using GPS and an unscented Kalman filter. *J. Geophys. Res.*, 114, B02405. doi: 10.1029/2008JB005837.

Freymueller, J.T., Murray, J.B., Rymer, H., Locke, C.A. (2015). Ground deformation, gravity, and magnetics. In *The Encyclopedia of Volcanoes*, Sigurdsson, H., Houghton, B., McNutt, S., Rymer, H., Stix, J. (eds). Elsevier, Amsterdam.

Gonnermann, H.M., Foster, J.H., Poland, M., Wolfe, C.J., Brooks, B.A., Miklius, A. (2012). Coupling at Mauna Loa and Kilauea by stress transfer in an asthenospheric melt layer. *Nature Geoscience*, 5(11), 826–829.

Gregg, P.M. and Pettijohn, J.C. (2016). A multi-data stream assimilation framework for the assessment of volcanic unrest. *Journal of Volcanology and Geothermal Research*, 309, 63–77.

Gudmundsson, M.T., Jónsdóttir, K., Hooper, A., Holohan, E.P., Halldórsson, S.A., Ófeigsson, B.G., Cesca, S., Vogfjörd, K.S., Sigmundsson, F., Högnadóttir, T. et al. (2016). Gradual caldera collapse at Bárdarbunga volcano, Iceland, regulated by lateral magma outflow. *Science*, 353(6296), aaf8988. doi: 10.1126/science.aaf8988.

Guldstrand, F., Galland, O., Hallot, E., Burchardt, S. (2018). Experimental constraints on forecasting the location of volcanic eruptions from pre-eruptive surface deformation. *Frontiers in Earth Science*, 6, 7.

Hooper, A., Zebker, H., Segall, P., Kampes, B. (2004). A new method for measuring deformation on volcanoes and other natural terrains using InSAR persistent scatterers. *Geophys. Res. Lett.*, 31, L23611. doi: 10.1029/2004GL021737.

Jaupart, C. and Allègre, C. (1991). Gas content, eruption rate and instabilities of eruption regime in silicic volcanoes. *Earth and Planetary Science Letters*, 102, 413–429.

Le Mével, H., Gregg, P.M., Feigl, K.L. (2016). Magma injection into a long-lived reservoir to explain geodetically measured uplift: Application to the 2007–2014 unrest episode at Laguna del Maule volcanic field, Chile. *Journal of Geophysical Research: Solid Earth*, 121(8), 6092–6108.

Lengliné, O., Marsan, D., Got, J.L., The, V., Ferrazzini, V., Okubo, P.G. (2008). Seismicity and deformation induced by magma accumulation at three basaltic volcanoes. *J. Geophys. Res. Solid Earth*, 113, 1–12.

Marzocchi, W., Sandri, L., Selva, J. (2008). Bet_ef: A probabilistic tool for long- and short-term eruption forecasting. *Bulletin of Volcanology*, 70(5), 623–632.

Massonnet, D. and Feigl, K.L. (1998). Radar interferometry and its application to changes in the Earth's surface. *Reviews of Geophysics*, 36(4), 441–500.

McTigue, D. (1987). Elastic stress and deformation near a finite spherical magma body: Resolution of the point source paradox. *Journal of Geophysical Research: Solid Earth*, 92(B12), 12931–12940.

Mériaux, C. and Jaupart, C. (1995). Simple fluid dynamic models of volcanic rift zones. *Earth and Planetary Science Letters*, 136(3/4), 223–240.

Michaut, C. (2011). Dynamics of magmatic intrusions in the upper crust: Theory and applications to laccoliths on earth and the moon. *J. Geophys. Res.*, 116, B05205. doi: 10.1029/2010JB008108.

Mogi, K. (1958). Relations between the eruptions of various volcanoes and the deformations of the ground surfaces around them. *Bulletin of the Earthquake Research Institute University of Tokyo*, 36, 99–134.

Perissin, D. and Ferretti, A. (2007). Urban-target recognition by means of repeated spaceborne sar images. *IEEE Transactions on Geoscience and Remote Sensing*, 45(12), 4043–4058.

Pinel, V. and Jaupart, C. (2003). Magma chamber behavior beneath a volcanic edifice. *J. Geophys. Res.*, 108(B2), 2072. doi: 10.1029/2002JB001751.

Pinel, V., Poland, M.P., Hooper, A. (2014). Volcanology: Lessons learned from synthetic aperture radar imagery. *Journal of Volcanology and Geothermal Research*, 289, 81–113.

Poland, M.P. and Anderson, K.R. (2020). Partly cloudy with a chance of lava flows: Forecasting volcanic eruptions in the twenty-first century. *Journal of Geophysical Research: Solid Earth*, 125(1), e2018JB016974.

Poland, M.P., Miklius, A., Sutton, A.J., Thornber, C.R. (2012). A mantle-driven surge in magma suply to Kilauea Volcano during 2003-2007. *Nature Geoscience*, 5, 295–300.

Reverso, T., Vandemeulebrouck, J., Jouanne, F., Pinel, V., Villemin, T., Sturkell, E., Bascou, P. (2014). A two-magma chamber model as a source of deformation at Grímsvötn Volcano, Iceland. *J. Geophys. Res. Solid Earth*, 119, 4666–4683.

Rivalta, E. and Segall, P. (2008). Magma compressibility and the missing source for some dike intrusions. *Geophysical Research Letters*, 35, L04306. doi: 10.1029/2007GL032521.

Rivalta, E., Corbi, F., Passarelli, L., Acocella, V., Davis, T., Di Vito, M.A. (2019). Stress inversions to forecast magma pathways and eruptive vent location. *Sci. Adv.*, 5, eaau9784.

Samiei-Esfahany, S. and Hanssen, R.F. (2013). New algorithm for InSAR stack phase triangulation using integer least squares estimation. In *2013 IEEE International Geoscience and Remote Sensing Symposium - IGARSS*, Melbourne.

Segall, P. (2013). Volcano deformation and eruption forecasting. *Geological Society, London, Special Publications*, 380(1), 85–106.

Shirzaei, M. and Walter, T. (2010). Time-dependent volcano source monitoring using interferometric synthetic aperture radar time series: A combined genetic algorithm and Kalman filter approach. *J. Geophys. Res.*, 115, B10421. doi: 10.1029/2010JB007476.

Sigmundsson, F., Hooper, A., Hreinsdóttir, S., Vogfjörd, K.S., Ófeigsson, B.G., Heimisson, E.R., Dumont, S., Parks, M., Spaans, K., Gudmundsson, G.B. et al. (2015). Segmented lateral dyke growth in a rifting event at Bardarbunga volcanic system, Iceland. *Nature*, 517(7533), 191–195.

Sigmundsson, F., Pinel, V., Grapenthin, R., Hooper, A., Halldórsson, S.A., Einarsson, P., Ófeigsson, B.G., Heimisson, E.R., Jónsdóttir, K., Gudmundsson, M.T. et al. (2020). Unexpected large eruptions from buoyant magma bodies within viscoelastic crust. *Nature Communications*, 11(1), 2403.

Sparks, R. (2003). Forecasting volcanic eruptions. *Earth and Planetary Science Letters*, 210(1), 1–15.

Sparks, R., Biggs, J., Neuberg, J. (2012). Monitoring volcanoes. *Science*, 335(6074), 1310–1311.

Wong, Y.-Q., Segall, P., Bradley, A., Anderson, K. (2017). Constraining the magmatic system at Mount St. Helens (2004–2008). Using Bayesian inversion with physics-based models including gas escape and crystallization. *Journal of Geophysical Research: Solid Earth*, 122(10), 7789–7812.

Wright, T.J., Parsons, B.E., Lu, Z. (2004). Toward mapping surface deformation in three dimensions using insar. *Geophys. Res. Lett.*, 31, L01607. doi: 10.1029/2003GL018827.

Xue, X., Freymueller, J., Lu, Z. (2020). Modeling the posteruptive deformation at Okmok based on the GPS and InSAR time series: Changes in the shallow magma storage system. *Journal of Geophysical Research: Solid Earth*, 125(2), e2019JB017801.

Yan, Y., Trouvé, E., Pinel, V., Mauris, G., Pathier, E., Galichet, S. (2012). Fusion of D-InSAR and sub-pixel image correlation measurements for coseismic displacement field estimation: Application to the Kashmir earthquake (2005). *International Journal of Image and Data Fusion*, 3(1), 71–92.

Yang, X.-M., Davis, P.M., Dieterich, J.H. (1988). Deformation from inflation of a dipping finite prolate spheroid in an elastic half-space as a model for volcanic stressing. *Journal of Geophysical Research: Solid Earth*, 93(B5), 4249–4257.

Zhan, Y. and Gregg, P.M. (2017). Data assimilation strategies for volcano geodesy. *Journal of Volcanology and Geothermal Research*, 344, 13–25.

Zhan, Y., Gregg, P.M., Chaussard, E., Aoki, Y. (2017). Sequential assimilation of volcanic monitoring data to quantify eruption potential: Application to Kerinci volcano, Sumatra. *Frontiers in Earth Science*, 5, 108.

Zhang, Z., Fattahi, H., Amelung, F. (2019). Small baseline InSAR time series analysis: Unwrapping error correction and noise reduction. *Computers & Geosciences*, 133, 104331. doi: 10.1016/j.cageo.2019.104331.

3

Data Assimilation in Glaciology

Fabien GILLET-CHAULET
IGE, Université Grenoble Alpes, France

3.1. Introduction

In its broadest sense, glaciology may combine a wide range of sciences that address natural phenomena and more generally involving ice. We will restrict ourselves here to studies concerning the polar ice sheets, and in particular those that focus on the deformation and flow of these ice masses.

Glaciers are masses of ice formed by the accumulation of snow, which becomes denser through compaction and metamorphism. At large time scales, ice behaves like a non-Newtonian viscous fluid and deforms under its own weight, consequently transporting ice to ablation zones. When these ice masses are large enough to cover the topography, these are then referred to as ice caps or ice sheets, the latter covering areas larger than 50,000 km^2.

Due to the long lifespan of ice sheets, their evolution is an integrated response to climate change. However, they are not just passive indicators of climate change, and important feedbacks take place with the other components of the climate system, atmosphere and oceans, and the solid earth (Figure 3.1). Since the early 1990s, the

---

For a color version of all the figures in this chapter, see www.iste.co.uk/yan/remotesensing.zip.

development of numerical models of the polar ice sheets has been driven by the need to constrain major climate changes and their feedbacks. In the current context, forecasting their short- and medium-term evolution has become an important societal issue, particularly in terms of their contribution to the global mean sea level (section 3.2). The general principles of model operation are described in section 3.3. With the increase in satellite observations over the last few decades, data assimilation methods have undergone significant development, with the aim of better constraining the geophysical processes at work in situ and better reproducing the changes observed to initialize projections. These methods and the questions they seek to address are presented in sections 3.4 and 3.5.

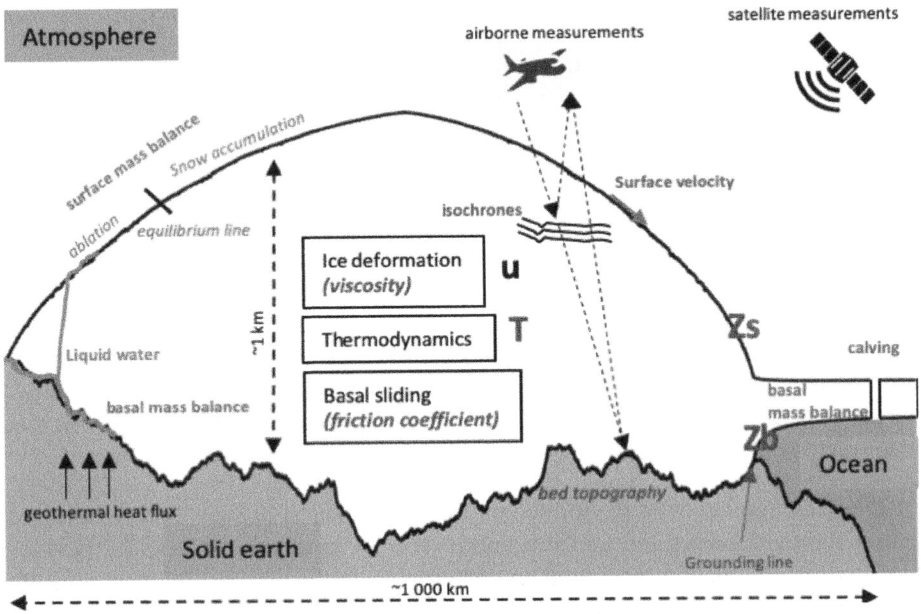

Figure 3.1. *Typical elements of a polar ice-sheet model. The ice sheet interacts with the atmosphere, the ocean and the solid earth*

COMMENT ON FIGURE 3.1.– *The polar ice sheet exchanges mass with the atmosphere and the ocean. The velocity field* $\mathbf{u}$ *is a diagnostic variable that depends on ice deformation and sliding at the base. The prognostic variables are the lower* Z_b *and upper* Z_s *free surfaces and the temperature field* T*. The most uncertain parameters are shown in purple. Satellite and airborne observations are used to measure surface velocities, surface altitude, bedrock topography and internal layers. The representation has been vertically dilated, because ice sheets are thin layers* *where typical thicknesses are significantly less than the characteristic horizontal dimensions.*

3.2. Predicting a paradigm shift for polar ice-sheet models

Numerical ice-sheet models emerged relatively recently, in the early 1990s for the first models capable of simulating the evolution of Antarctica, Greenland or the now extinct ice sheets (Huybrechts et al. 1996). Used with a spatial resolution and physics particularly well suited to representing ice flow in the core areas of the ice sheets, they have led to a view in which the ice sheets were a slow component of the climate system gradually responding to paleo-climatic changes.

Paradoxically, although the first hypotheses concerning the stability of marine ice sheets, namely those resting on a bedrock below sea level as illustrated in Figure 3.1, date back to the 1970s (Weertman 1974; Mercer 1978), this view was only questioned when the first observations of fast changes at the margins of the ice sheets appeared. In 2007, the IPCC's fourth report on climate change mentioned that the processes affecting ice dynamics revealed by observations were still missing from the models (IPCC 2007). In 2013, in the fifth report (Church et al. 2013), changes in glacier dynamics were still assessed independently from the literature rather than through physical modeling. These findings prompted the development of new-generation polar sheet models (Gillet-Chaulet et al. 2012) and a paradigm shift in modeling polar sheets (Pattyn 2018). Answering the question "How much and how fast can the polar sheets contribute to sea-level change?" has become a major challenge for the community (Scambos et al. 2017).

Recently, the ISMIP6 ice-sheet model intercomparison project has brought the community together to provide physical projections of Antarctica's and Greenland's contribution to sea level (Goelzer et al. 2020; Seroussi et al. 2020). This project is an integral part of the CMIP6 coupled model intercomparison project (Eyring et al. 2016). For the most pessimistic evolution of greenhouse gas concentrations (RCP8.5), these studies predict a contribution to mean sea level in 2100 of 9 ± 5 cm for Greenland (Goelzer et al. 2020) and ranging from -7.8 to +30 cm for Antarctica (Seroussi et al. 2020).

Compared with other climate system components, the polar ice sheets may appear as a fairly simple system to be numerically modeled. They are composed of a single material that viscously deforms under its own weight. The flow is so slow that turbulence, Coriolis forces and other inertial effects can be largely ignored. It is especially in the prescription of boundary conditions at the base of the cap and at the margins that things get complicated (Vaughan and Arthern 2007). Despite major improvements in model sophistication, there is still quite a difference in the processes they include, the spatial resolution of their underlying grids and the forcing that they apply (Goelzer et al. 2017; Pattyn et al. 2017). This is partly due to the variety of questions they address: the same "model" can be used for process studies at the glacier or watershed scale over a few years or decades, or even for continental-scale simulations covering a few hundred or thousand years.

In addition to solved physics, the efficiency of predictive models also relies heavily on the quality of the data available for validation and testing. Unlike weather forecasting models, which may have to deal with observations on a daily basis and can be improved when they prove inadequate, ice-sheet models cannot rely as much on this validation cycle due to the smaller amount of data available and the longer time scales involved (Vaughan and Arthern 2007). The spatial and temporal resolution of observations has presently considerably increased, especially by satellites (Figure 3.2), which makes it possible to assess the state of the mass balance of ice sheets and glaciers, as well as their sensitivity to ongoing climate change. Nonetheless, this still represents a relatively short period in terms of climate and for the assessment of century-scale projections (Hanna et al. 2020).

For these projections, the initial state has been identified as a major source of uncertainty (Ađalgeirsdóttir et al. 2014), and the first inter-comparisons carried out in ISMP6 aimed to compare and evaluate the initialization methods of the community in order to better estimate the sources of uncertainty linked to the initial state (Goelzer et al. 2018; Seroussi et al. 2019). In general, two different approaches can be distinguished:

– Since a large number of models have been first developed for paleo-climatological studies, the initial state is constructed from a *spin-up* of the model over long time periods, ranging from the millennia to severals of millennia. This results in obtaining a consistent state for long-term response processes such as thermodynamics or isostasy (Goelzer et al. 2017; Pattyn et al. 2017). Therefore, the initial state for short-term projections is in fact the final state of a long paleo-climatic simulation, and this state may differ significantly from observations, particularly in terms of thickness and spatial extent. To overcome this problem, and thus approximate the topography that is actually under observation, some models implemented *nudging* methods (Rückamp et al. 2019) or iterative methods where certain model parameters, often friction at the base, were corrected using rules which often were heuristic (Pollard and DeConto 2012; Pattyn 2017; Le clec'h et al. 2019). Although somewhat resembling data assimilation, we will not describe these methods in detail here.

– The other conventional approach consists of using the observed topography and assimilating surface velocity observations. We describe these methods in section 3.3.

In order to maximize spatial resolution, the fields used are generally derived from observations averaged over several years. Given that the solved equations of mechanics are non-inertial, the inverse problem is tantamount to one of parameter estimation. These methods therefore did not capture observed trends (Goldberg and Heimbach 2013). Moreover, due to residual uncertainties, models initialized in this way generally showed flow divergence anomalies that led to artificial model drift for simulations starting from these states (Seroussi et al. 2011). To overcome this problem, the topography was usually relaxed over a few years.

These intercomparisons showed that the difference in initial topography had a fairly strong influence on the volume changes predicted by the models (Goelzer et al. 2018; Seroussi et al. 2019). On the other hand, in regions where the simulated ice sheets overlap, there was good agreement between the models for the response of glacier dynamics to changes in surface mass balance, which are gradual changes. Conversely, for forcings that can induce instabilities, such as mass balance perturbations under floating platforms in Antarctica, there remains considerable variation in model response. These variations are considered to be due to differences in initial topography, with modeled platform extents that can quite widely vary between models, but also to initial ocean conditions and numerical processing of the grounding line (Seroussi et al. 2019).

Although the observation period is still relatively short, the satellite observation boom has allowed comparing models to increasingly dense time series. For example, Aschwanden et al. (2013) showed that comparing Greenland's historical simulations from 1989 to 2011 with available time series could already lead to reducing the number of acceptable initial states. Indeed, initial states in good agreement with spatially dense observations of surface velocity and change in altitude at the initial state may misrepresent other important characteristics, such as ice flux to the ocean or ice thickness, and thus result in different evolutions. Data assimilation methods that can be used to take the transient nature of observations and model dynamics into account are gradually being developed. These methods will be presented in section 3.5.

3.3. Principles of ice sheet dynamics

Over long-term periods, ice behaves like a non-Newtonian, incompressible viscous fluid. The flow of ice is governed by Stokes equations, which express the conservation of mass and momentum:

$$\begin{cases} \mathrm{div}\boldsymbol{u} = 0 \\ \mathrm{div}\boldsymbol{\sigma} + \rho\boldsymbol{g} = 0 \end{cases} \quad [3.1]$$

where $\boldsymbol{u}$ is the velocity vector, ρ is the ice density, $\boldsymbol{g}$ is the gravity vector, $\boldsymbol{\sigma} = \boldsymbol{\tau} - p\boldsymbol{I}$ is the Cauchy stress tensor, with $\boldsymbol{\tau}$ the deviatoric stress tensor, $p = -\mathrm{tr}(\boldsymbol{\sigma})/3$ the isotropic pressure and $\boldsymbol{I}$ the identity matrix.

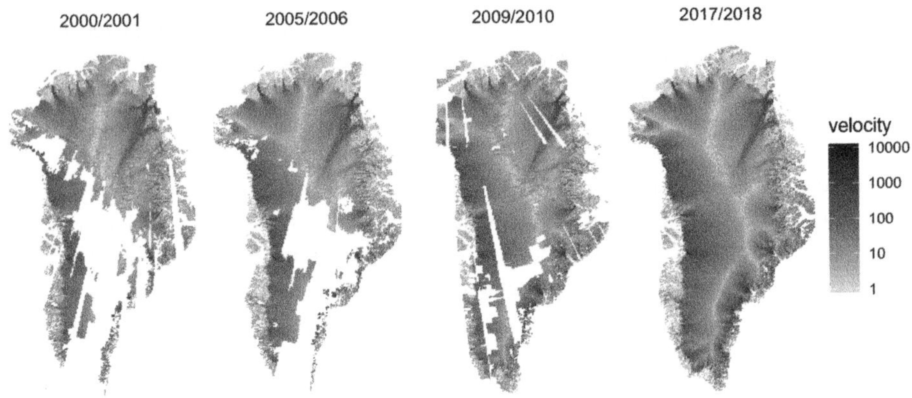

Figure 3.2. *Winter surface velocities for the Greenland ice sheet measured by synthetic aperture radar (SAR) showing increasing spatial coverage between winters 2000/2001, 2005/2006, 2009/2010 and 2017/2018 (Joughin et al. 2010). The velocity field is very heterogeneous with just 15 outflow glaciers responsible for half of the ice flow discharged into the ocean each year (Enderlin et al. 2014)*

This three-dimensional equation system of unknowns $\boldsymbol{u}$ and p is closed by adopting a flow law for the ice material that relates the deviatoric stress tensor to the strain rate tensor $\dot{\boldsymbol{\epsilon}}$. Most models adopt a nonlinear Norton–Hoff-type behavior law known as Glen's law in glaciology (Glen 1955):

$$\boldsymbol{\tau} = 2\eta\dot{\boldsymbol{\epsilon}} \qquad [3.2]$$

where the effective viscosity η is defined by:

$$\eta = \frac{1}{2}(EA)^{-1/n}\dot{\epsilon}_e^{(1-n)/n} \qquad [3.3]$$

where $\dot{\epsilon}_e^2 = \mathrm{tr}(\dot{\boldsymbol{\epsilon}}^2)/2$ is the square of the second invariant of strain rates, $A(T')$ is a rheological parameter that depends on T', the temperature of the ice relative to the pressure melting point and E an ad hoc amplification factor generally used to account for uncertainty in this flow law. Within an ice sheet, stresses and strain rates vary over several orders of magnitude, making the processes responsible for in situ deformation particularly difficult to reproduce in the laboratory. There is therefore considerable uncertainty about the form of the law (equation [3.2]) and the associated parameters (Alley 1992; Gillet-Chaulet et al. 2011).

Due to the low aspect ratios of the ice sheets (Figure 3.1), several thin-film approximations to the Stokes equations have been developed, significantly reducing numerical costs. The most conventional in large-scale models are the *shallow ice approximation* (SIA, Hutter et al. 1981), and the *shelfy stream approximation* (SSA, (MacAyeal 1989)); however, several other so-called higher order approximations have been developed and implemented (e.g. Schoof and Hindmarsh 2010). The SIA is valid at the center of the ice sheets, where the ice is frozen on its base. This approximation does not consider longitudinal stresses compared to vertical shear stresses, which depend on surface slope and ice thickness. As its name implies, the SSA is valid for floating platforms and ice streams where friction at the base is low compared to other constraints. The result is a vertically integrated, two-dimensional equation system with horizontal velocity components as the two unknowns.

The natural boundary conditions for a glacier are those at the interfaces with the atmosphere, ocean and bedrock. At atmospheric and oceanic interfaces, a Neumann condition on normal stress is applied:

$$\boldsymbol{\sigma}.\boldsymbol{n} = p_{a/w}\boldsymbol{n} \qquad [3.4]$$

with $p_a \approx 0$ as the negligible pressure of the atmosphere and p_w as the pressure due to seawater.

It is at the bedrock level that conditions become particularly uncertain. When ice is cold, it is generally assumed that it adheres perfectly to its bedrock, leading to zero sliding velocity. In areas where the ice is temperate and allows the presence of liquid water, the ice slide on its base can contribute significantly to the dynamics and velocities observed at the surface. The condition at the base is then generally given by a friction law that links the basal stress tangent to the base $\boldsymbol{\tau_b} = \boldsymbol{T}.\boldsymbol{\sigma}.\boldsymbol{n}$ to the sliding velocity $\boldsymbol{u_b} = \boldsymbol{T}.\boldsymbol{u}$:

$$\boldsymbol{\tau_b} + \beta\boldsymbol{u_b} = \boldsymbol{0} \qquad [3.5]$$

where the operator $\boldsymbol{T} = \boldsymbol{I} - \boldsymbol{n} \otimes \boldsymbol{n}$ allows the projection onto the tangent plane to the base with normal $\boldsymbol{n}$. The coefficient β is an effective coefficient of friction that will generally depend on velocity, bedrock properties and the pressure difference between ice and subglacial water. The development of a general friction law for representing sliding at the base of the polar ice sheets is one of the long-standing problems of glaciology (Schoof 2005; Zoet and Iverson 2020).

Since equations [3.1] are non-inertial, the velocity field depends solely on geometry, material properties and boundary conditions. Therefore, a significant number of studies have focused on estimating uncertain model parameters, material properties and boundary conditions, from observations of surface velocities, with surface topography imposed from the observations. These methods will be presented in section 3.4.

Time dependency is reflected in the prognostic variables of the model. In general, equations [3.1] are coupled to a thermodynamics module (Greve and Blatter 2016). This module is used to compute the temperature field on which ice viscosity depends (equation [3.2]) and makes it possible to distinguish temperate areas, where ice can slide on its base, from cold areas. It is the long response time of thermodynamics that requires particularly long *spin-ups* to build up the initial states of ice sheets, as the temperature field still keeps the trace of conditions at the last glacial maximum. Nevertheless, for models initialized by inverse methods, the 100-year simulations are not really responsive to the thermal regime (Seroussi et al. 2013).

Finally, this is a free-surface problem, and the models solve a mass conservation equation for the evolution of the topography of the cap. For vertically integrated models, this translates into an equation for the evolution of ice thickness H:

$$\frac{\partial H}{\partial t} + \mathrm{div}(\boldsymbol{u}H) = \dot{a} \qquad [3.6]$$

with $\dot{a}$ an accumulation/ablation function representing mass exchanges with the atmosphere and ocean. This forcing is generally prescribed from climate models and represents one of the major sources of uncertainty in projections. Vertically integrated models assume that the pressure p is hydrostatic, and the altitude of free surfaces in contact with the atmosphere and ocean is directly deduced from H and the bedrock topography by evaluating the buoyancy condition. Three-dimensional models must solve the continuity equations for the upper and lower free surfaces separately. For the lower surface, the position of the grounding line is the solution to the contact problem between the lower surface and the bedrock, which adds significant numerical complexity (Gagliardini et al. 2013). At long timescales, the evolution of subglacial topography cannot be neglected, and models include an isostasy module.

3.4. Parameter estimation

The first inverse methods appeared fairly quickly with the first satellite observations of ice-sheet surface velocities (MacAyeal 1992). Since the velocity field at a given time t is a diagnostic variable in the mechanical equations governing flow (equation [3.1]), it is possible to formulate an inverse problem that aims to estimate the model parameter fields, in particular, the basal friction coefficient β (equation [3.5]) and ice viscosity η (equation [3.2]), from surface observations. Here, we present the methods that are most widely used in the community to initialize flow models, and then discuss the most common applications and issues.

3.4.1. *Variational methods*

3.4.1.1. *Control method*

Many models implement a method derived from control theory and applied as early as 1992 to the reconstruction of frictional conditions at the base of an Antarctic ice stream (MacAyeal 1992). Detailed by MacAyeal (1993) for an SSA model, it has been adapted to most classical approximations (Goldberg and Sergienko 2011; Morlighem et al. 2013) and the full Stokes equations (Morlighem et al. 2010; Gillet-Chaulet et al. 2012; Petra et al. 2012).

The problem then amounts to a nonlinear least-squares problem, where the optimal parameters sought for are those that minimize a cost function J measuring the deviation between model solutions and observations. Conventionally, when the observations are the horizontal components of the surface velocities, with $\boldsymbol{u}^{obs}$ measured on the upper surface Γ_s, the cost function is defined as:

$$J_c = \int_{\Gamma_s} \frac{1}{2} |\boldsymbol{u} - \boldsymbol{u}^{obs}|^2 ds \qquad [3.7]$$

Given that the parameter field searched for is generally high dimensional, minimization is based on a gradient method, and therefore requires computing the derivatives of J_c with respect to the parameters searched for. In order to compute these gradients, J_c is augmented using Lagrange multipliers to add the constraint that $\boldsymbol{u}$ is a solution to equations [3.1]. The minimization conditions for the Lagrange functional lead to the expression of the adjoint model and gradients with respect to the desired parameters.

By neglecting the non-linearity of viscosity (equation [3.2]) and friction (equation [3.5]), the problem is self-adjoint and the analytical expression for the adjoint model can be obtained quite simply. For the Stokes equations, the adjoint model is given by (Gagliardini et al. 2013):

$$\begin{cases} 2\text{div}(\eta\dot{\epsilon}^{\lambda}) - \text{grad}q = 0 \\ \text{tr}(\dot{\epsilon}^{\lambda}) = 0 \end{cases} \qquad [3.8]$$

where the variables λ and q correspond to the *adjoint* velocity and pressure variables, respectively, and $\dot{\epsilon}^{\lambda}$ is the equivalent of the strain tensor for the variable λ.

The boundary conditions of this system of equations are then given by the derivatives of the cost function J_c (equation [3.7]). This introduces great flexibility in the formulation of J_c by, for example, comparing only the norm of velocities (Gillet-Chaulet et al. 2012) or adding a term corresponding to the relative error (Morlighem et al. 2010).

The Gateaux derivatives of J_c with respect to β and η in directions β' and η', respectively, are given by:

$$\begin{array}{l} d_{\eta}J_c(\eta,\beta,\eta') = \int_{\Omega} -2\eta' \left(\dot{\epsilon}^{\lambda}.\dot{\epsilon}\right) d\Omega \\ d_{\beta}J_c(\eta,\beta,\beta') = \int_{\Gamma_b} -\beta' \boldsymbol{u}.\boldsymbol{\lambda} d\Gamma \end{array} \qquad [3.9]$$

with Ω the domain and Γ_b the basal boundary.

The fact that the system is self-adjoint and the expressions can be analytically obtained has greatly facilitated the implementation of these methods, which are now seen as the most conventional. Furthermore, Petra et al. (2012) showed that for a nonlinear model (whether for viscosity or friction law), the adjoint operator of the Stokes equations is the same as the Jacobian that appears when the direct (nonlinear) problem was solved by a Newton method. Thereby, any direct code including a Newton method for the Stokes model was already equipped with the operator needed for solving the adjoint model. Exact expressions for nonlinear models were also derived for several approximations of the Stokes equations (Goldberg and Sergienko 2011; Morlighem et al. 2013).

It should be noted that the adjoint code could also be obtained by automatic differentiation of the source code (Goldberg and Heimbach 2013; Martin and Monnier 2014). For their part, Brinkerhoff and Johnson (2013) and Kyrke-Smith et al. (2017) employed the dolphin-adjoint library, which automatically led to automatically deriving the adjoint code from the symbolic representation, in variational form, of the equations to be solved; the implementation of the direct and adjoint codes was carried out automatically by the FEniCS finite element software (Farrell et al. 2013).

For a code based on vertically integrated mechanical equations, ice thickness, and therefore subglacial topography, appeared as a model parameter, and the derivative of the cost function could be obtained in the same way (Perego et al. 2014; Mosbeux et al. 2016). For three-dimensional models, thickness is not a parameter but is directly given by the shape of the domain. The gradient of J_c with respect to subglacial topography was given by Perego et al. (2014) for a Stokes model with a vertically structured mesh.

3.4.1.2. *The Robin inverse method*

To solve the same type of problem, Arthern and Gudmundsson (2010) adapted a method originally developed for acoustic impedance tomography (Chaabane and Jaoua 1999). The β coefficient of friction of the basal friction law (equation [3.5]) was seen as an indeterminate coefficient for a Robin boundary condition, that is, a linear combination of Dirichlet and Neumann boundary conditions.

The inverse problem was translated into a variational problem where the solution minimized a particular form of cost function introduced by Kohn and Vogelius (1984). Compared with the control methods previously presented, the advantage of this method is that it does not require the development of a specific solver. In fact, the method consists of alternately solving: (i) the classical direct Neumann-type problem for which the natural stress condition is applied at the level of the upper surface, and (ii) a Dirichlet-type problem where surface velocity observations are applied as Dirichlet conditions.

Denoting by ^{N} and ^{D} the solutions to the Neumann and Dirichlet problems, respectively, the optimal viscosity and basal friction coefficient fields minimize the following cost function:

$$J_R(\eta, \beta) = \int_\Omega 2\eta \left| \dot{\epsilon}^N - \dot{\epsilon}^D \right|_F^2 + \int_{\Gamma_b} \beta \left| \boldsymbol{u}^N - \boldsymbol{u}^D \right|^2 = \int_{\Gamma_s} (\boldsymbol{u}^N - \boldsymbol{u}^D).(\boldsymbol{\sigma}^N - \boldsymbol{\sigma}^D).\boldsymbol{n} \quad [3.10]$$

where $|_F$ denotes the Frobenius norm.

Considering linear behavior and friction laws, Arthern and Gudmundsson (2010) analytically derived the gradient of the cost function J_R (equation [3.10]) with respect to β and η. The Gateaux derivatives of J_R are given by:

$$\begin{aligned} d_\eta J_R(\eta, \beta, \eta') &= \textstyle\int_\Omega 2\eta' \left(|\dot{\epsilon}^D|_F^2 - |\dot{\epsilon}^N|_F^2 \right) d\Omega \\ d_\beta J_R(\eta, \beta, \beta') &= \textstyle\int_{\Gamma_b} \beta' \left(|\boldsymbol{u}^D|^2 - |\boldsymbol{u}^N|^2 \right) d\Gamma \end{aligned} \quad [3.11]$$

The method was extended to a vertically integrated model by Arthern et al. (2015), who used the continuity equation for ice thickness (equation [3.6]) as an additional constraint for the Dirichlet model.

3.4.1.3. *Validation, minimization and regularization*

The developed methods were generally validated with twin experiments, where the aim was to recover model parameters from synthetic observations generated with the direct model (MacAyeal 1993; Arthern and Gudmundsson 2010). However, for variational methods, the accuracy of the cost function gradient calculation could be evaluated by comparison with the gradient computed by finite difference for a given perturbation direction (Gagliardini et al. 2013; Martin and Monnier 2014). When nonlinear terms are overlooked, the gradients obtained with the control methods (equation [3.9]) and Robin (equation [3.11]) led to similar accuracies, limited by the assumption on nonlinearity (Gagliardini et al. 2013). The methods were compared by Gillet-Chaulet et al. (2012) for inversion of the basal friction field with a *full-Stokes* model of Greenland and very similar results were obtained. The advantages of the control method over the Robin method is therefore mainly associated with the flexibility provided for defining the cost function and, in particular, incorporating observation errors; and with the fact that the equations presented here can be extended with the control method to take account of nonlinearities in the behavior or friction law.

The impact of nonlinearity was studied in more detail by Martin and Monnier (2014). In their model, the nonlinearity of the behavior law was resolved by an iterative Picard method. In the adjoint code obtained by automatic differentiation, this was achieved by inverse accumulation, where the iterations carried out by the direct code had to be performed in reverse order by the adjoint code. Neglecting nonlinearities with the adjoint model amounts to performing only the first iteration of the adjoint code. They showed that the accuracy of the gradients is significantly reduced with this assumption compared with the full solution, and that this could limit the wavelengths identifiable when basal conditions were inverted from surface observations.

Similarly, from synthetic experiments with a vertically integrated model, Goldberg and Sergienko (2011) showed that keeping nonlinear terms affects the ability of the minimization algorithm to find the minimum independently of initialization and led to better convergence. On the other hand, for realistic applications, they did not find any significant difference, probably due to uncertainties in the observations and other model parameters. This result was confirmed by Morlighem et al. (2013), who obtained similar convergences when preserving or overlooking nonlinear terms for inversions of basal conditions with a three-dimensional higher order model of Antarctica.

Once the gradient has been calculated, efficiency is highly dependent on the minimization algorithm. For the Robin method, Arthern and Gudmundsson (2010) proposed a higher gradient descent algorithm for minimizing J_R by taking spatially uniform descent steps. Schäfer et al. (2012) and Arthern et al. (2015), following a heuristic method, proposed a spatially variable descent step that allowed better convergence for realistic applications where parameters and speeds could vary over several orders of magnitude. Incomplete Gauss–Newton methods were developed by

Habermann et al. (2012) and Petra et al. (2012), which showed that they performed significantly better than conventional greater slope or nonlinear conjugate gradient methods. Alternatively, several models utilized L-BFGS-based quasi-Newton methods implemented in open-access libraries (Gilbert and Lemaréchal 1989).

Despite the fact that improved computational resources and methods made it possible to apply these methods with models that fully solved the Stokes equations at Greenland's (Gillet-Chaulet et al. 2012) and Antarctica's scales (Petra et al. 2012), this remained numerically costly and many models solved more or less advanced approximations. The difference in response of direct models was addressed in several model inter-comparison exercises (Pattyn et al. 2008, 2012), but the error resulting from the approximation remained unconstrained for realistic applications. Concerning the results of inverse methods, based on analytical solutions, Gudmundsson (2008) compared the transmission of perturbations in the basal conditions to the upper free surface for the Stokes model and conventional approximations. He showed in particular that SSA overestimates the surface transfer of short-wavelength perturbations of basal topography. This could affect inversions, in particular by underestimating the effects of uncertainties in the observations and overestimating the spatial resolution capabilities obtained when inversions of surface observations were performed. Therefore, with a twin experiment where surface observations were generated with a Stokes model, Habermann et al. (2017) showed that, in fast flowing areas, the residual errors obtained by performing the inversion with an SSA model were of the same order of magnitude as those obtained for inversions with real data. More precisely, Morlighem et al. (2010) showed that inversions conducted with approximations of the Stokes equations could lead to solutions different from those obtained with the full model in order to compensate for neglected constraints in the approximated solutions.

As in other fields, inverse problems in glaciology are poorly posed, and regularization must be introduced in order to obtain a stable solution. With variational methods, the minimization stopping criterion can be used as a regularization technique. Indeed, iterative methods tend to correct large-scale features first, and the minimization stopping criterion thus influences the scales present in the inverse parameters (Habermann et al. 2012). An ad hoc stopping criterion based on the improvement of the cost function between successive iterations was used, for example, in the first realistic applications (Rommelaere and MacAyeal 1997; Joughin et al. 2004). Since this criterion is difficult to relate to observation and model errors, Vieli and Payne (2003) proposed a stopping criterion based on the misfit principle, where minimization was conducted until the cost function reached a certain tolerance that could be constrained from observation errors. From synthetic tests, Arthern and Gudmundsson (2010) and Habermann et al. (2012) showed that if iterations proceeded once this criterion was exceeded, the inverted parameters tended to show erroneous, high-amplitude, high-frequency characteristics. For realistic applications where model and observation errors are poorly constrained, it may be difficult to define a priori such a criterion.

Another classic regularization method consists of adding terms to the cost function to be minimized. Therefore, most models use a Tikhonov regularization where the additional terms penalize a given norm of the inverted parameters. A conventional choice in glaciology consists of penalizing the first spatial derivatives of the inverse parameters in order to prefer smooth solutions. When a background can be estimated for the parameter to be inverted, some models used a Sobolev H1 norm, possibly supplemented by an L2 norm measuring the deviation from the background (e.g. Habermann et al. 2013). These regularization functions generally need to be weighted by a regularization parameter. When there is only one parameter to calibrate, the most commonly used method is the L-curve method (e.g. Jay-Allemand et al. 2011), where the optimum parameter corresponds to the maximal curvature when representing the evolution of the regularization function according to the error term. Things get more complicated when there are several regularization parameters to calibrate, which is often done on a trial-and-error basis using the principles of the L-curve as a baseline and trying to find a good trade-off between smooth solutions and deviation from observations (e.g. Fürst et al. 2015).

The need to calibrate these regularization weights, which are not always reported in the literature, makes it difficult to compare the results of different studies with each other. Using a Tikhonov regularization is tantamount to introducing a certain spatial correlation structure into the parameters searched for (Arthern 2015; Brinkerhoff et al. 2016). Prescribing information a priori in the form of background and a background error covariance matrix is not very usual in glaciology. Firstly, certainly because of difficulties for its implementation, but also due to the difficulty in constraining the terms to be introduced, particularly for basal conditions which are so poorly defined a priori. In Isaac et al. (2015), regularization was introduced in the form of a diffusion operator, which, for an unstructured mesh such as is conventionally used in glaciology models, amounted to imposing a background error covariance matrix where spatial correlations were given by Matern-type correlation functions (Guillet et al. 2019). Using a simplified case study, Arthern (2015) discussed the possibility of using a group transformation principle and a relative entropy maximization method that served to more formally prescribe the probability density function a priori. He showed that this complicated the problem but would result in a more rational and less subjective formulation of the regularization terms.

3.4.2. *Bayesian methods*

The same types of problem have been addressed using Bayesian methods. Therefore, Gudmundsson and Raymond (2008) were mainly concerned with the reconstruction of basal friction and topography from surface topography and velocities. To relate basal conditions to surface observations, they used a linear model in the form of analytically derived transfer functions (Gudmundsson 2003). This made it possible to analytically derive the maximum a posteriori (MAP) estimator from the a priori and their error covariance matrix. This model was extended for a direct nonlinear model by Raymond and Gudmundsson (2009) and applied by Pralong and Gudmundsson (2011). The optimal solution then corresponded to the minimum of a cost function whose derivatives were approximated using the transfer functions of Gudmundsson (2003).

Still assuming an a posteriori Gaussian probability function, Isaac et al. (2015) proposed an efficient algorithm for finding the MAP. The method was applied to the inversion of the friction coefficient field for a 3D model of Antarctica, where the total number of parameters to be estimated was about 10^6. For problems of this size, it proved impossible to evaluate the covariance matrix a posteriori and represent it by an operator; they proposed a low-rank approximation of this matrix in order to estimate the uncertainty of the inverted field and propagate this uncertainty in future simulations.

Petra et al. (2012) moved away from the Gaussian assumption and proposed a Markov chain Monte Carlo method. To guide sampling toward higher probability regions, they used locally Gaussian approximations based on low-rank approximations of the Hessian matrix. The method is applied to friction field inversion for a 2D model representing a flow line giving just over 100 parameters to be inverted.

3.4.3. *Classical inversion problems*

3.4.3.1. *Basal friction*

Ice sheets outlet glaciers drain a large proportion of the ice flow to the margins and are therefore important for the mass balance of the ice sheets (Enderlin et al. 2014). In these fast-flowing areas, slip at the base is a major contributor to the velocities observed at the surface. Since conditions at the base of the ice sheets are, by definition, not directly accessible for measuring, the most conventional inverse problem consists of estimating the basal-friction field β (equation [3.5]) from the surface velocities (MacAyeal 1992). These methods are currently the only ones that can be used to obtain spatial estimates of basal stress and slip velocity beneath the ice sheets, which are the fields needed for testing more complex models of the subglacial environment (Arthern 2015).

In general, inversions showed that the bedrock beneath these outlets provides little resistance to flow and supports a very small fraction of the driving stress generated by the surface slope. (MacAyeal 1992; Joughin et al. 2004; Morlighem et al. 2010; Shapero et al. 2016). However, most of these studies mentioned the presence of areas where resistance is much greater, thus potentially playing an important role in flow stability. For Antarctica, Arthern et al. (2015) showed that the usual estimate in glaciology that basal stress was of the order of a hundred kilopascals was founded at first sight; nonetheless, it could vary locally over several orders of magnitude and friction at the base of the ice sheets could not be simply represented by a coefficient of friction or a spatially uniform stress. The inverted friction field for the Greenland ice sheet is shown in Figure 3.3. Fast flow areas are generally characterized by low friction. In general, the effective friction coefficient β decreases with velocity to give a constant stress between 10 kPa and 1 MPa.

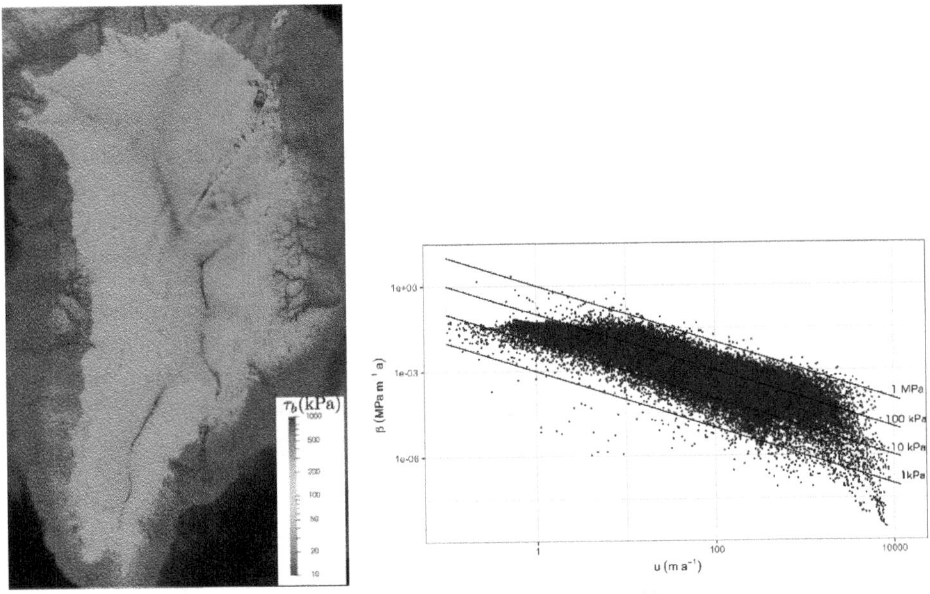

a) b)

Figure 3.3. *Assimilation results for surface velocity observations to reverse friction at the base of Greenland with the Elmer/Ice model (Gagliardini et al. 2013)*

COMMENT ON FIGURE 3.3.– *(a) The basal stress τ_b and (b) the relation between the friction coefficient β and the sliding speed u_b inverted. The black lines are iso values of the stress τ_b following relation [3.5]. It is generally not possible to obtain a simple parameterization, and these results are used to improve understanding of the processes controlling the sliding of the ice on its base.*

An important question concerns the robustness of inversions to recover and really constrain spatial variations in basal conditions. On a synthetic case, Gudmundsson and Raymond (2008) investigated the limits of spatial resolution and of the information that can be obtained from surface observations. Furthermore, they found that, for typical measurement errors, small-amplitude perturbations in basal friction could only be resolved at wavelengths greater than 50 times the average ice thickness. In addition, the inverted basal friction is highly sensitive to errors in basal topography. This particular aspect was studied in more detail by Kyrke-Smith et al. (2018). Variations in topography induce form resistance to flow. Therefore, a significant fraction of the inverted stress may be due to insufficient resolution of the subglacial topography. In contrast, the spatial structure of inverted fields is considered to be fairly robust, but highly sensitive to regularization. Sensitivity to regularization was discussed in several other studies (Habermann et al. 2012; Shapero et al. 2016); however, as explained previously, since most models are based on Tikhonov regularization, which rather favors smooth fields, results are difficult to compare from one study to another.

The sensitivity of reconstructions to other sources of model uncertainty such as mesh resolution, temperature field and surface topography was addressed in several other studies (Joughin et al. 2004; Schäfer et al. 2012; Habermann et al. 2017). In particular, Habermann et al. (2017) showed that these other sources of uncertainty can have a significant effect on the inverted friction field and therefore on its interpretation. On the other hand, these uncertainties lead to spatial characteristics of characteristic residual error and could be used to improve models or orientate data acquisition. One way of improving these reconstructions would be to use other data sources such as acoustic impedance measurements; however, despite the fact that there exists some level of correlation between these measurements and inverted friction fields, theoretical knowledge was still too poor to incorporate this information (Kyrke-Smith et al. 2017).

The inverted basal stress field must satisfy the global equilibrium of the ice sheet. For a surface observation field acquired at a single given date, the inverted field is therefore largely unaffected by the choice of the form of the friction law used (Joughin et al. 2004). On the other hand, projections are highly affected by this friction law (Ritz et al. 2015) and, with increasing temporal resolution of surface observations, several studies focused on the temporal variability of basal friction (Jay-Allemand et al. 2011; Habermann et al. 2013; Minchew et al. 2016; Gilbert et al. 2018). Theories show that friction at the base depends on subglacial water pressure. In areas where surface melting is significant, subglacial hydrology is strongly influenced by water coming from the surface and therefore modulates friction. The conventional approach is therefore to carry out a series of independent inversions with different velocity fields and interpret the temporal variations of the inverted friction fields. For an area in Antarctica where surface melting is still negligible and over a period of time where other basal conditions can be assumed to be constant, Gillet-Chaulet et al. (2016) formulated a slightly different inverse problem with the objective of finding a spatially varying, but time-constant friction coefficient that best reproduces the velocity fields observed between 1996 and 2010. The best agreement with observations was obtained with highly nonlinear friction laws, showing that the bedrock deforms in a quasi-plastic manner with a stress that depends very little on velocity. The bedrock is therefore not able to take up the excess stresses resulting from the loss of resistive stresses at the margins in contact with the ocean, making the watershed highly sensitive to perturbations.

3.4.3.2. *Viscosity*

The ice shelves surrounding Antarctica play an important role in the stability of the ice sheets. Confined to bays or laying on islets, they exert a buttressing effect and hold back the flow of ice laid upstream. For example, the outflow glaciers of the Larsen B platform, which suddenly disappeared in 2002, saw their speed increase by a factor of 2–6 between 2000 and 2003 (Scambos 2004). Sun et al. (2020) evaluated the response of the ice cap to an extreme scenario that would see the total disappearance of all platforms. The global mean sea level could then rise by 1–12 m over a period of 500 years.

For these platforms, friction at the base in contact with the ocean is zero, and the main unknown is ice rheology. The SSA equations were developed to describe the flow of these platforms, and the most classic inverse problem involves inverting a two-dimensional mean viscosity field from surface observations (Rommelaere and MacAyeal 1997). Inversions generally show high variability in this viscosity, explained by factors such as cold ice advection from tributary glaciers, basal melting and the presence of severely damaged shear zones or rifts. Inversions achieved based on time series of surface velocities have contributed to better understanding and constraining the mechanisms, leading to their weakening and potential destabilization (Khazendar et al. 2007; Borstad et al. 2013; Khazendar et al. 2015). As with friction, inversions are quite affected by residual errors in the model. For example, based on inversions achieved over the whole of the Antarctic, Fürst et al. (2015) showed that areas where model velocities did not agree with observations corresponded to pinning points not referenced in the reference topography used for inversions. The results made it possible to estimate the buttressing effect of *ice shelfs*, showing that only a small percentage of the shelf area had a negligible role in the balance of forces (Fürst et al. 2016). Thereby, any perturbation of these platforms, such as their thinning as a result of increased melting at the base in contact with the ocean, could affect the current balance and thus modify ice flows originating from the continent.

3.4.3.3. *Base topography*

Subglacial topography is generally measured by airborne radar campaigns. Given the areas to be covered, knowledge of the relief beneath the ice sheets is thus very patchy, and reference topographies for Greenland and Antarctica report uncertainties of several hundred meters for areas not covered (Morlighem et al. 2017, 2020). As mentioned in the methods previously presented, surface velocities are impacted by the sub-glacial topography, which could thus be integrated into the control parameter space (Gudmundsson and Raymond 2008; Perego et al. 2014; Mosbeux et al. 2016).

The other conventional approach for constraining subglacial topography consists of formulating an inverse problem from the continuity equation for ice thickness (equation [3.6]), which guarantees the conservation of ice fluxes. As such, Morlighem et al. (2013) derived the adjoint model from equation [3.6]; the control parameter space was then given by mean velocities and surface-mass balance, and for the purpose of reproducing single-point thickness observations. This method is particularly suitable for ice streams, where the observed surface velocities are a good approximation of the mean velocities and have a low relative uncertainty. This method is now commonly used at the margins of the ice sheets to build reference topographies (Morlighem et al. 2017, 2020). This inverse problem was also formulated within a Bayesian framework by Brinkerhoff et al. (2016).

Smaller sized continental glaciers also represent an important reservoir of freshwater, essential for many populations, and currently contribute significantly to sea-level rise. Direct measurements of ice thickness are only available for a very small percentage of the total number of glaciers. As a result, methods for reconstructing ice volumes and subglacial topography from surface observations have seen considerable development in recent years. Classical approaches were recently compared in an inter-comparison exercise which included 17 models (Farinotti et al. 2017). The methods are quite varied, depending in particular on the intended application, with some models aiming to be applied to all glaciers, which generally requires a high degree of automation and simpler assumptions. The inter-comparison did not reveal any model performing better than the others, and in general, the solution given by the average of all the models is better than each model individually considered, which is expected for an unbiased ensemble. All approaches are based on stationarity assumptions, and methods that use more observations (surface velocities and topography, surface mass balance) do not outperform simpler methods that use only surface topography, probably because of inconsistencies and uncertainties in the datasets.

3.4.3.4. *Inversion of several parameters*

Most of the studies cited above only invert one parameter field at a time, and assume that the other model parameters are known, which often does not correspond to reality. Several studies have therefore investigated the possibility of simultaneously inverting several fields.

For the simultaneous inversion of basal topography and friction, Raymond and Gudmundsson (2009) noted that the effects of these two fields could be fairly well separated from surface observations, and that the risk of mixing between the two fields was not a serious problem. Nonetheless, they observed that basal friction inversion was very sensitive to errors in the topography, probably because the roughness of the topography induced shape constraints that were compensated for by friction when the topography utilized was too smooth (Kyrke-Smith et al. 2018). Goldberg and Heimbach (2013) noted, however, that inversion of the two fields was more accurate when the two fields are spatially uncorrelated. On twin experiments, Mosbeux et al. (2016) also managed to separate friction and topography in fast-flowing areas quite well, notably by adding a constraint based on the continuity equation (equation [3.6]) in addition to surface velocity observations.

The results are less clear for the simultaneous inversion of viscosity and basal friction. On a simple analytical case, Arthern and Gudmundsson (2010) showed that there could be an infinite number of solutions for reproducing surface velocities if no a priori constraints were imposed on the two parameters. Simultaneous inversion of the two fields is now standard practice for initializing projections made with ice-sheet models, but the robustness of these inversions is not generally discussed in detail. With a Tikhonov regularization that gives preference to smooth solutions, it is difficult to consistently impose an a priori and its uncertainty that would allow the problem to be scaled. Arthern et al. (2015) and Brondex et al. (2019) addressed this issue and performed several inversions, giving more or less weight to one of the two fields during minimization. The admissible solutions were then selected by analyzing the results obtained and estimating whether they were compatible with the uncertainties generally accepted in the literature.

3.5. State and parameter estimation

In recent years, the development of new space missions has considerably contributed to increase the spatial and temporal resolution of observations. One of the challenges for models is to develop methods to assimilate these large masses of data, with their very diverse properties in terms of resolution and accuracy (Pattyn and Morlighem 2020).

Transient assimilation methods for complete models including velocity computations and free surface evolution have so far only been applied to synthetic cases or regional studies (Goldberg and Heimbach 2013; Larour et al. 2014; Bonan et al. 2014; Goldberg et al. 2015; Gillet-Chaulet 2020).

Since many models are already equipped with variational methods and the adjoint model for computing the velocity field, one approach is to extend this type of method to the transient case and thus to 4D-Var assimilation. As we saw in the previous section, the adjoint model of the momentum equations is generally obtained analytically and can be coded relatively simply. This becomes impractical for transient models, and the use of automatic source code differentiation software becomes a necessity. The methodological and technical difficulties specific to ice-sheet models have been addressed in several recent studies. In particular, the most important computational cost in models often derives from solving the system of nonlinear elliptic equations for computing the velocity field (equation [3.1]). Nonlinearity is usually solved by inverse accumulation (Martin and Monnier 2014). Goldberg et al. (2016) implemented an efficient algorithm that required replacing a small part of the automatically generated code with hand-written code. As this only affected a small part of the code, it did not interfere with the modularity offered by automatic differentiation and does not require revision as the physics of the model changes. On the other hand, fairly recent codes often use languages and structures not supported by existing software. As a result, most automatic differentiation tools have been written for Fortran code, and support for other languages remains limited. Hascoët and Morlighem (2018) discussed the associated challenges in differentiating a C code clone from the ISSM code written in C++. Due to this particularity, ISSM also implements an adjoint model obtained with an operator overloading approach that is much more costly in terms of memory (Larour et al. 2014, 2016).

These methods could be used for real applications for Greenland and Antarctic outflow glaciers. Larour et al. (2014) assimilated annual surface elevation observations over six years for a Greenland outlet glacier by constraining basal friction and surface mass balance. As their observations did not cover an entire drainage basin, Goldberg et al. (2015) restricted their study to the grounded part of an outlet glacier in West Antarctica, and assimilated surface elevation and velocities over a decade to constrain conditions at the domain boundary. The model trajectory was in good agreement with observations, and the initial state was used to predict the mass loss already underway in this region. Finally, at the scale of entire ice sheets, Heimbach and Bugnion (2009) and Logan et al. (2020) applied automatic differentiation software to obtain the adjoint code of the SICOPOLIS model. The resulting codes were not used for data assimilation, but the adjoint code was used to compute the sensitivity of ice-sheet mass to different model parameters.

Requiring less development effort, ensemble methods, based in particular on the ensemble Kalman filter, have been successfully applied for data assimilation in many areas of geophysics, even with nonlinear codes and a large number of unknowns. These methods have only recently been tested in glaciology and only with synthetic cases. Therefore, Bonan et al. (2014) implemented an *ensemble transform Kalman filter* (ETKF) (Hunt et al. 2007) with a model based on SIA approximations. The performance of the method was studied for a synthetic twin experiment. Annual observations of surface velocity and altitude obtained by perturbing the output of a direct model simulation were used to constrain basal topography and friction. They showed that these two fields could be fairly well constrained for realistic levels of uncertainty in surface observations, even for relatively small ensembles using localization and inflation techniques.

The method was extended in order to be able to assimilate the position of the margins in a model that explicitly tracks this boundary with an adaptive mesh method. Gillet-Chaulet (2020) used the *error subspace transform Kalman filter* (ESTKF) (Nerger et al. 2012) implemented in the PDAF library (Nerger et al. 2005) with a model based on SSA equations. The method was applied to a synthetic case of a marine ice sheet that underwent instability, causing a significant retreat of the grounding line. The assimilation of annual observations of surface velocity and altitude enables basal topography and friction to be well constrained. By correcting the state, represented by ice thickness, the method was also able to constrain the evolution of the grounding line over the assimilation window. On the other hand, since the stability of the grounding line is highly dependent on basal conditions, residual uncertainties can have a significant impact on projections. The results showed that retreat rates were properly reproduced by the model, but that it was very difficult to predict the exact chronology of the retreat.

3.6. Conclusions and outlook

With the observation of rapid changes at the margins of the ice sheets and their significant contribution to sea level rise in recent decades, polar ice-sheet modeling has entered a new era. Models have undergone significant advances in solved physics and numerical methods. However, the variety of questions addressed by these models, and the spatial and temporal scales involved, implies that the approaches remain highly variable. As such, there is still no consensus on the procedures for initializing models to produce 100-year projections of the contribution of ice sheets to sea level.

The development of satellite missions in recent years has significantly led to improved observational coverage and spatial resolution; and on the other hand, the period considered remains short for model evaluation. Most models have implemented inverse methods that enable parameters to be calibrated from surface velocity observations. Despite these methods being commonly used to initialize models and attempt to better constrain unobservable processes such as friction at the base of the ice, further work is needed to unambiguously characterize the current state of the ice sheets. Indeed, as currently implemented, regularization methods involve a large amount of subjectivity as they affect the spatial scales that can be inferred from surface observations and the separation between the inverted parameters.

Often static, conventional methods prove unable to capture trends currently observed. One of the challenges for the coming years will be to develop methods robust enough for consuming data recently acquired and assimilating them in a time-consistent manner. Promising methods have emerged in recent years, but have yet to prove themselves in real-life applications. One of the difficulties is that the main changes are taking place at the margins of the ice sheets and in outlet glaciers, which drain most of ice fluxes, and methods developed for applications dedicated to a particular glacier are unlikely to be transposable to the global scale.

Another challenge will be to integrate other sources of information than those conventionally utilized presently. For current changes, the observations which are not yet in widespread use are the positions of calving fronts and grounding lines. These data correspond to the limits of the domain and will require appropriate developments to be assimilated. Since the period of satellite observation is relatively short compared with the response time of certain processes, it would be interesting to be able to go further back into the past in order to put current changes into perspective with wider climate changes. The internal layers of the ice sheets, which are measured by penetrating radar at the same time as the sub-glacial topography, are still rarely used. These layers are isochronous and therefore represent a measure of past surface states that have been advected and deformed by flow. The use of this information would undoubtedly result in better constraining changes over longer time scales.

Another major limitation of current methods is that they generally lead to a single initial state, the uncertainty of which is very difficult to estimate, and how it is reflected in the projections. Future developments should therefore enable not only that initial states could be improved by assimilating more and better observations, but also by better constraining the associated uncertainty.

3.7. References

Ađalgeirsdóttir, G., Aschwanden, A., Khroulev, C., Boberg, F., Mottram, R., Lucas-Picher, P., Christensen, J.H. (2014). Role of model initialization for projections of 21st-century Greenland ice sheet mass loss. *J. Glaciol.*, 60(222), 782–794.

Alley, R.B. (1992). Flow-law hypotheses for ice-sheet modeling. *J. Glaciol.*, 38(129), 245–256.

Arthern, R.J. (2015). Exploring the use of transformation group priors and the method of maximum relative entropy for Bayesian glaciological inversions. *J. Glaciol.*, 61(229), 947–962.

Arthern, R.J. and Gudmundsson, G.H. (2010). Initialization of ice-sheet forecasts viewed as an inverse Robin problem. *J. Glaciol.*, 56(197), 527–533.

Arthern, R.J., Hindmarsh, R.C.A., Williams, C.R. (2015). Flow speed within the Antarctic ice sheet and its controls inferred from satellite observations. *J. Geophys. Res., Earth Surf.*, 120(7), 1171–1188.

Aschwanden, A., Aðalgeirsdóttir, G., Khroulev, C. (2013). Hindcasting to measure ice sheet model sensitivity to initial states. *The Cryosphere*, 7(4), 1083–1093.

Bonan, B., Nodet, M., Ritz, C., Peyaud, V. (2014). An ETKF approach for initial state and parameter estimation in ice sheet modelling. *Nonlinear Processes in Geophysics*, 21(2), 569–582.

Borstad, C.P., Rignot, E., Mouginot, J., Schodlok, M.P. (2013). Creep deformation and buttressing capacity of damaged ice shelves: Theory and application to Larsen C ice shelf. *The Cryosphere*, 7(6), 1931–1947.

Brinkerhoff, D.J. and Johnson, J.V. (2013). Data assimilation and prognostic whole ice sheet modelling with the variationally derived, higher order, open source, and fully parallel ice sheet model VarGlaS. *The Cryosphere*, 7(4), 1161–1184.

Brinkerhoff, D.J., Aschwanden, A., Truffer, M. (2016). Bayesian inference of subglacial topography using mass conservation. *Front. Earth Sci.*, 4. doi: 10.3389/feart.2016.00008.

Brondex, J., Gillet-Chaulet, F., Gagliardini, O. (2019). Sensitivity of centennial mass loss projections of the Amundsen basin to the friction law. *The Cryosphere*, 13(1), 177–195.

Chaabane, S. and Jaoua, M. (1999). Identification of Robin coefficients by the means of boundary measurements. *Inverse Probl.*, 15(6), 1425–1438.

Church, J.A., Clark, P.U., Cazenave, A., Gregory, J.M., Jevrejeva, S., Levermann, A., Merrifield, M.A., Milne, G.A., Nerem, R.S., Nunn, P.D. et al. (2013). Sea level change. In *Climate Change 2013: The Physical Science Basis. Contribution of Working Group I to the Fifth Assessment Report of the Intergovernmental Panel on Climate Change*, Stocker, T.F., Qin, D., Plattner, G.-K. et al. (eds). Cambridge University Press, Cambridge, 1137–1216.

Enderlin, E.M., Howat, I.M., Jeong, S., Noh, M.-J., Angelen, J., Broeke, M.R. (2014). An improved mass budget for the Greenland ice sheet. *Geophys. Res. Lett.*, 41(3), 866–872.

Eyring, V., Bony, S., Meehl, G.A., Senior, C.A., Stevens, B., Stouffer, R.J., Taylor, K.E. (2016). Overview of the Coupled Model Intercomparison Project Phase 6 (CMIP6) experimental design and organization. *Geosci. Model Dev.*, 9(5), 1937–1958.

Farinotti, D., Brinkerhoff, D.J., Clarke, G.K., Fürst, J.J., Frey, H., Gantayat, P., Gillet-Chaulet, F., Girard, C., Huss, Leclercq, P.W. et al. (2017). How accurate are estimates of glacier ice thickness? Results from ITMIX, the Ice Thickness Models Intercomparison eXperiment. *The Cryosphere*, 11(2), 949–970.

Farrell, P.E., Ham, D.A., Funke, S.W., Rognes, M.E. (2013). Automated derivation of the adjoint of high-level transient finite element programs. *SIAM J. Sci. Comput.*, 35(4), C369–C393.

Fürst, J.J., Durand, G., Gillet-Chaulet, F., Merino, N., Tavard, L., Mouginot, J., Gourmelen, N., Gagliardini, O. (2015). Assimilation of Antarctic velocity observations provides evidence for uncharted pinning points. *The Cryosphere*, 9(4), 1427–1443.

Fürst, J.J., Durand, G., Gillet-Chaulet, F., Tavard, L., Rankl, M., Braun, M., Gagliardini, O. (2016). The safety band of Antarctic ice shelves. *Nature Clim. Change*, 6, 479–482.

Gagliardini, O., Zwinger, T., Gillet-Chaulet, F., Durand, G., Favier, L., de Fleurian, B., Greve, R., Malinen, M., Martín, C., Råback, P. et al. (2013). Capabilities and performance of Elmer/Ice, a new-generation ice sheet model. *Geosci. Model Dev.*, 6(4), 1299–1318.

Gilbert, J.C. and Lemaréchal, C. (1989). Some numerical experiments with variable-storage quasi-Newton algorithms. *Mathematical Programming*, 45(1), 407–435.

Gilbert, A., Leinss, S., Kargel, J., Kääb, A., Gascoin, S., Leonard, G., Berthier, E., Karki, A., Yao, T. (2018). Mechanisms leading to the 2016 giant twin glacier collapses, Aru Range, Tibet. *The Cryosphere*, 12(9), 2883–2900.

Gillet-Chaulet, F. (2020). Assimilation of surface observations in a transient marine ice sheet model using an ensemble Kalman filter. *The Cryosphere*, 14(3), 811–832.

Gillet-Chaulet, F., Hindmarsh, R.C.A., Corr, H.F.J., King, E.C., Jenkins, A. (2011). In-situ quantification of ice rheology and direct measurement of the Raymond effect at Summit, Greenland using a phase-sensitive radar. *Geophys. Res. Lett.*, 38, 6.

Gillet-Chaulet, F., Gagliardini, O., Seddik, H., Nodet, M., Durand, G., Ritz, C., Zwinger, T., Greve, R., Vaughan, D.G. (2012). Greenland ice sheet contribution to sea-level rise from a new-generation ice-sheet model. *The Cryosphere*, 6(6), 1561–1576.

Gillet-Chaulet, F., Durand, G., Gagliardini, O., Mosbeux, C., Mouginot, J., Rémy, F., Ritz, C. (2016). Assimilation of surface velocities acquired between 1996 and 2010 to constrain the form of the basal friction law under Pine Island Glacier. *Geophys. Res. Lett.*, 43(19), 2016GL069937.

Glen, J.W. (1955). The creep of polycrystalline ice. *Proc. R. Soc. Lond. A*, 228(1175), 519–538.

Goelzer, H., Robinson, A., Seroussi, H., van de Wal, R.S. (2017). Recent progress in Greenland ice sheet modelling. *Current Climate Change Reports*, 3(4), 291–302.

Goelzer, H., Nowicki, S., Edwards, T., Beckley, M., Abe-Ouchi, A., Aschwanden, A., Calov, R., Gagliardini, O., Gillet-Chaulet, F., Golledge, N.R. (2018). Design and results of the ice sheet model initialisation experiments initMIP-Greenland: An ISMIP6 intercomparison. *The Cryosphere*, 12(4), 1433–1460.

Goelzer, H., Nowicki, S., Payne, A., Larour, E., Seroussi, H., Lipscomb, W.H., Gregory, J., Abe-Ouchi, A., Shepherd, A. (2020). The future sea-level contribution of the Greenland ice sheet: A multi-model ensemble study of ISMIP6. *The Cryosphere*, 14(9), 3071–3096.

Goldberg, D.N. and Heimbach, P. (2013). Parameter and state estimation with a time-dependent adjoint marine ice sheet model. *The Cryosphere*, 7(6), 1659–1678.

Goldberg, D.N. and Sergienko, O.V. (2011). Data assimilation using a hybrid ice flow model. *The Cryosphere*, 5(2), 315–327.

Goldberg, D.N., Heimbach, P., Joughin, I., Smith, B. (2015). Committed retreat of Smith, Pope, and Kohler Glaciers over the next 30 years inferred by transient model calibration. *The Cryosphere*, 9(6), 2429–2446.

Goldberg, D.N., Narayanan, S.H.K., Hascoet, L., Utke, J. (2016). An optimized treatment for algorithmic differentiation of an important glaciological fixed-point problem. *Geosci. Model Dev.*, 9(5), 1891–1904.

Greve, R. and Blatter, H. (2016). Comparison of thermodynamics solvers in the polythermal ice sheet model SICOPOLIS. *Polar Science*, 10(1), 11–23.

Gudmundsson, G.H. (2003). Transmission of basal variability to a glacier surface. *J. Geophys. Res.*, 108(B5), 2253. doi: 10.1029/2002JB002107.

Gudmundsson, G.H. (2008). Analytical solutions for the surface response to small amplitude perturbations in boundary data in the shallow-ice-stream approximation. *The Cryosphere*, 2(2), 77–93.

Gudmundsson, G.H. and Raymond, M. (2008). On the limit to resolution and information on basal properties obtainable from surface data on ice streams. *The Cryosphere*, 2(2), 167–178.

Guillet, O., Weaver, A.T., Vasseur, X., Michel, Y., Gratton, S., Gürol, S. (2019). Modelling spatially correlated observation errors in variational data assimilation using a diffusion operator on an unstructured mesh. *Quart. J. Roy. Meteor. Soc.*, 145(722), 1947–1967.

Habermann, M., Maxwell, D., Truffer, M. (2012). Reconstruction of basal properties in ice sheets using iterative inverse methods. *J. Glaciol.*, 58(210), 795–808.

Habermann, M., Truffer, M., Maxwell, D. (2013). Changing basal conditions during the speed-up of Jakobshavn Isbræ, Greenland. *Cryosphere*, 7(6), 1679–1692.

Habermann, M., Truffer, M., Maxwell, D. (2017). Error sources in basal yield stress inversions for Jakobshavn Isbræ, Greenland, derived from residual patterns of misfit to observations. *J. Glaciol.*, 63(242), 999–1011.

Hanna, E., Pattyn, F., Navarro, F., Favier, V., Goelzer, H., van den Broeke, M.R., Vizcaino, M., Whitehouse, P.L., Ritz, C., Bulthuis, K. et al. (2020). Mass balance of the ice sheets and glaciers – Progress since AR5 and challenges. *Earth-Science Reviews*, 201, 102976.

Hascoët, L. and Morlighem, M. (2018). Source-to-source adjoint Algorithmic Differentiation of an ice sheet model written in C. *Optimization Methods and Software*, 33(4/6), 829–843.

Heimbach, P. and Bugnion, V. (2009). Greenland ice-sheet volume sensitivity to basal, surface and initial conditions derived from an adjoint model. *Ann. Glaciol.*, 50(52), 67–80.

Hunt, B.R., Kostelich, E.J., Szunyogh, I. (2007). Efficient data assimilation for spatiotemporal chaos: A local ensemble transform Kalman filter. *Physica D: Nonlinear Phenomena*, 230(1/2), 112–126.

Hutter, K., Legerer, F., Spring, U. (1981). First-order stresses and deformations in glaciers and ice sheets. *J. Glaciol.*, 27(96), 227–270.

Huybrechts, P., Payne, T., Group, T.E.I. (1996). The EISMINT benchmarks for testing ice-sheet models. *Ann. Glaciol.*, 23, 1–12.

Isaac, T., Petra, N., Stadler, G., Ghattas, O. (2015). Scalable and efficient algorithms for the propagation of uncertainty from data through inference to prediction for large-scale problems, with application to flow of the Antarctic ice sheet. *J. Comput. Phys.*, 296, 348–368.

Jay-Allemand, M., Gillet-Chaulet, F., Gagliardini, O., Nodet, M. (2011). Investigating changes in basal conditions of Variegated Glacier prior to and during its 1982–1983 surge. *The Cryosphere*, 5(3), 659–672.

Joughin, I., MacAyeal, D.R., Tulaczyk, S. (2004). Basal shear stress of the Ross ice streams from control method inversions. *J. Geophys. Res.*, 109(B9), B09405.

Joughin, I., Smith, B.E., Howat, I.M., Scambos, T., Moon, T. (2010). Greenland flow variability from ice-sheet-wide velocity mapping. *J. Glaciol.*, 56(197), 415–430.

Khazendar, A., Rignot, E., Larour, E. (2007). Larsen B Ice Shelf rheology preceding its disintegration inferred by a control method. *Geophys. Res. Lett.*, 34(19), L19503. doi: 10.1029/2007GL030980.

Khazendar, A., Borstad, C.P., Scheuchl, B., Rignot, E., Seroussi, H. (2015). The evolving instability of the remnant Larsen B Ice Shelf and its tributary glaciers. *Earth Planet. Sci. Lett.*, 419, 199–210.

Kohn, R. and Vogelius, M. (1984). Determining conductivity by boundary measurements. *Communications on Pure and Applied Mathematics*, 37(3), 289–298.

Kyrke-Smith, T.M., Gudmundsson, G.H., Farrell, P.E. (2017). Can seismic observations of bed conditions on ice streams help constrain parameters in ice flow models? *J. Geophys. Res., Earth Surf.*, 122(11), 2269–2282.

Kyrke-Smith, T.M., Gudmundsson, G.H., Farrell, P.E. (2018). Relevance of detail in basal topography for basal slipperiness inversions: A case study on Pine Island Glacier, Antarctica. *Front. Earth Sci.*, 6. doi: 10.3389/feart.2018.00033.

Larour, E., Utke, J., Csatho, B., Schenk, A., Seroussi, H., Morlighem, M., Rignot, E., Schlegel, N., Khazendar, A. (2014). Inferred basal friction and surface mass balance of the Northeast Greenland Ice Stream using data assimilation of ICESat (Ice Cloud and land Elevation Satellite) surface altimetry and ISSM (Ice Sheet System Model). *The Cryosphere*, 8(6), 2335–2351.

Larour, E., Utke, J., Bovin, A., Morlighem, M., Perez, G. (2016). An approach to computing discrete adjoints for MPI-parallelized models applied to Ice Sheet System Model 4.11. *Geosci. Model Dev.*, 9(11), 3907–3918.

Le clec'h, S., Quiquet, A., Charbit, S., Dumas, C., Kageyama, M., Ritz, C. (2019). A rapidly converging initialisation method to simulate the present-day Greenland ice sheet using the GRISLI ice sheet model (version 1.3). *Geoscientific Model Development*, 12(6), 2481–2499.

Logan, L.C., Narayanan, S.H.K., Greve, R., Heimbach, P. (2020). SICOPOLIS-AD v1: An open-source adjoint modeling framework for ice sheet simulation enabled by the algorithmic differentiation tool OpenAD. *Geoscientific Model Development*, 13(4), 1845–1864.

MacAyeal, D.R. (1989). Large-scale ice flow over a viscous basal sediment: Theory and application to ice stream B, Antarctica. *J. Geophys. Res.*, 94(B4), 4071–4087.

MacAyeal, D.R. (1992). The basal stress distribution of Ice Stream E, Antarctica, inferred by control methods. *J. Geophys. Res. Solid Earth*, 97(B1), 595–603.

MacAyeal, D.R. (1993). A tutorial on the use of control methods in ice-sheet modeling. *J. Glaciol*, 39(131), 91–98.

Martin, N. and Monnier, J. (2014). Adjoint accuracy for the full Stokes ice flow model: Limits to the transmission of basal friction variability to the surface. *The Cryosphere*, 8(2), 721–741.

Mercer, J.H. (1978). West Antarctic ice sheet and CO 2 greenhouse effect: A threat of disaster. *Nature*, 271(5643), 321–325.

Minchew, B., Simons, M., Björnsson, H., Palsson, F., Morlighem, M., Seroussi, H., Larour, E., Hensley, S. (2016). Plastic bed beneath Hofsjökull Ice Cap, central Iceland, and the sensitivity of ice flow to surface meltwater flux. *J. Glaciol.*, 62(231), 147–158.

Morlighem, M., Rignot, E., Seroussi, H., Larour, E., Dhia, H.B., Aubry, D. (2010). Spatial patterns of basal drag inferred using control methods from a full-Stokes and simpler models for Pine Island Glacier, West Antarctica. *Geophys. Res. Lett.*, 37, 6.

Morlighem, M., Seroussi, H., Larour, E., Rignot, E. (2013). Inversion of basal friction in Antarctica using exact and incomplete adjoints of a higher-order model. *J. Geophys. Res., Earth Surf.*, 118(3), 1746–1753.

Morlighem, M., Williams, C.N., Rignot, E., An, L., Arndt, J.E., Bamber, J.L., Catania, G., Chauché, N., Dowdeswell, J.A., Dorschel, B. et al. (2017). BedMachine v3: Complete bed topography and ocean bathymetry mapping of Greenland from multibeam echo sounding combined with mass conservation. *Geophys. Res. Lett.*, 44(21), 11051–11061.

Morlighem, M., Rignot, E., Binder, T., Blankenship, D., Drews, R., Eagles, G., Eisen, O., Ferraccioli, F., Forsberg, R., Fretwell, P. et al. (2020). Deep glacial troughs and stabilizing ridges unveiled beneath the margins of the Antarctic ice sheet. *Nat. Geosci.*, 13(2), 132–137.

Mosbeux, C., Gillet-Chaulet, F., Gagliardini, O. (2016). Comparison of adjoint and nudging methods to initialise ice sheet model basal conditions. *Geosci. Model Dev.*, 9(7), 2549–2562.

Nerger, L., Hiller, W., Schröter, J. (2005). PDAF - The Parallel Data Assimilation Framework: Experiences with Kalman filtering. In *Use of High Performance Computing in Meteorology*, Zwieflhofer, W. and Mozdzynski, G. (eds). World Scientific, Reading.

Nerger, L., Schröter, J., Hiller, W. (2012). A unification of ensemble square root Kalman filters. *Monthly Weather Review*, 140(7), 2335–2345.

Pattyn, F. (2017). Sea-level response to melting of Antarctic ice shelves on multi-centennial timescales with the fast Elementary Thermomechanical Ice Sheet model (f.ETISh v1.0). *The Cryosphere*, 11(4), 1851–1878.

Pattyn, F. (2018). The paradigm shift in Antarctic ice sheet modelling. *Nat Commun*, 9(1), 1–3.

Pattyn, F. and Morlighem, M. (2020). The uncertain future of the Antarctic ice sheet. *Science*, 367(6484), 1331–1335.

Pattyn, F., Perichon, L., Aschwanden, A., Breuer, B., de Smedt, B., Gagliardini, O., Gudmundsson, G.H., Hindmarsh, R.C.A., Hubbard, A., Johnson, J.V. et al. (2008). Benchmark experiments for higher-order and full-Stokes ice sheet models (ISMIP–HOM). *The Cryosphere*, 2(2), 95–108.

Pattyn, F., Schoof, C., Perichon, L., Hindmarsh, R.C.A., Bueler, E., de Fleurian, B., Durand, G., Gagliardini, O., Gladstone, R., Goldberg, D. et al. (2012). Results of the Marine Ice Sheet Model Intercomparison Project, MISMIP. *The Cryosphere*, 6(3), 573–588.

Pattyn, F., Favier, L., Sun, S., Durand, G. (2017). Progress in numerical modeling of Antarctic ice-sheet dynamics. *Curr Clim Change Rep*, 3(3), 174–184.

Perego, M., Price, S., Stadler, G. (2014). Optimal initial conditions for coupling ice sheet models to Earth system models. *J. Geophys. Res. Earth Surf.*, 119, 1894–1917. doi: 10.1002/2014JF003181.

Petra, N., Zhu, H., Stadler, G., Hughes, T.J., Ghattas, O. (2012). An inexact Gauss-Newton method for inversion of basal sliding and rheology parameters in a nonlinear Stokes ice sheet model. *J. Glaciol.*, 58(211), 889–903.

Pollard, D. and DeConto, R.M. (2012). Description of a hybrid ice sheet-shelf model, and application to Antarctica. *Geosci. Model Dev.*, 5(5), 1273–1295.

Pralong, M.R. and Gudmundsson, G.H. (2011). Bayesian estimation of basal conditions on Rutford Ice Stream, West Antarctica, from surface data. *J. Glaciol.*, 57(202), 315.

Raymond, M.J. and Gudmundsson, G.H. (2009). Estimating basal properties of ice streams from surface measurements: A non-linear Bayesian inverse approach applied to synthetic data. *The Cryosphere*, 3(2), 265–278.

Ritz, C., Edwards, T.L., Durand, G., Payne, A.J., Peyaud, V., Hindmarsh, R.C.A. (2015). Potential sea-level rise from Antarctic ice-sheet instability constrained by observations. *Nature*, 528, 115–118.

Rommelaere, V. and MacAyeal, D.R. (1997). Large-scale rheology of the Ross Ice Shelf, Antarctica, computed by a control method. *Ann. Glaciol.*, 24, 43–48.

Rückamp, M., Greve, R., Humbert, A. (2019). Comparative simulations of the evolution of the Greenland ice sheet under simplified Paris Agreement scenarios with the models SICOPOLIS and ISSM. *Polar Science*, 21, 14–25.

Scambos, T.A. (2004). Glacier acceleration and thinning after ice shelf collapse in the Larsen B embayment, Antarctica. *Geophys. Res. Lett.*, 31(18), L18402. doi: 10.1029/2004GL020670.

Scambos, T.A., Bell, R.E., Alley, R.B., Anandakrishnan, S., Bromwich, D.H., Brunt, K., Christianson, K., Creyts, T., Das, S.B., DeConto, R. et al. (2017). How much, how fast? A science review and outlook for research on the instability of Antarctica's Thwaites Glacier in the 21st century. *Global and Planetary Change*, 153, 16–34.

Schäfer, M., Zwinger, T., Christoffersen, P., Gillet-Chaulet, F., Laakso, K., Pettersson, R., Pohjola, V.A., Strozzi, T., Moore, J.C. (2012). Sensitivity of basal conditions in an inverse model: Vestfonna ice cap, Nordaustlandet/Svalbard. *The Cryosphere*, 6(4), 771–783.

Schoof, C. (2005). The effect of cavitation on glacier sliding. *Proceedings of the Royal Society A: Mathematical, Physical and Engineering Sciences*, 461(2055), 609–627.

Schoof, C. and Hindmarsh, R.C.A. (2010). Thin-film flows with wall slip: An asymptotic analysis of higher order glacier flow models. *The Quarterly Journal of Mechanics and Applied Mathematics*, 63(1), 73–114.

Seroussi, H., Morlighem, M., Rignot, E., Larour, E., Aubry, D., Dhia, H.B., Kristensen, S.S. (2011). Ice flux divergence anomalies on 79north Glacier, Greenland. *Geophys. Res. Lett.*, 38, 5.

Seroussi, H., Morlighem, M., Rignot, E., Khazendar, A., Larour, E., Mouginot, J. (2013). Dependence of century-scale projections of the Greenland ice sheet on its thermal regime. *J. Glaciol.*, 59(218), 1024–1034.

Seroussi, H., Nowicki, S., Simon, E., Abe-Ouchi, A., Albrecht, T., Brondex, J., Cornford, S., Dumas, C., Gillet-Chaulet, F., Goelzer, H. et al. (2019). initMIP-Antarctica: An ice sheet model initialization experiment of ISMIP6. *The Cryosphere*, 13(5), 1441–1471.

Seroussi, H., Nowicki, S., Payne, A.J., Goelzer, H., Lipscomb, W.H., Abe-Ouchi, A., Agosta, C., Albrecht, T., Asay-Davis, X., Barthel, A. et al. (2020). ISMIP6 Antarctica: A multi-model ensemble of the Antarctic ice sheet evolution over the 21st century. *The Cryosphere*, 14(9), 3033–3070.

Shapero, D.R., Joughin, I.R., Poinar, K., Morlighem, M., Gillet-Chaulet, F. (2016). Basal resistance for three of the largest Greenland outlet glaciers. *J. Geophys. Res. Earth Surf.*, 121, 168–180. doi: 10.1002/2015JF003643.

Solomon, S., Qin, D., Manning, M., Chen, Z., Marquis, M., Averyt, K.B., Tignor, M., Miller, H.L. (2007). *Climate Change 2007: The Physical Science Basis. Contribution of Working Group I to the Fourth Assessment Report of the Intergovernmental Panel on Climate Change*. Cambridge University Press, Cambridge.

Sun, S., Pattyn, F., Simon, E.G., Albrecht, T., Cornford, S., Calov, R., Dumas, C., Gillet-Chaulet, F., Goelzer, H., Golledge, N.R. et al. (2020). Antarctic ice sheet response to sudden and sustained ice-shelf collapse (ABUMIP). *J. Glaciol.*, 66(260), 891–904.

Vaughan, D.G. and Arthern, R. (2007). Why is it hard to predict the future of ice sheets? *Science*, 315(5818), 1503–1504.

Vieli, A. and Payne, A.J. (2003). Application of control methods for modelling the flow of Pine Island Glacier, West Antarctica. *Annals Glaciol.*, 36(1), 197–204.

Weertman, J. (1974). Stability of the junction of an ice sheet and an ice shelf. *Journal of Glaciology*, 13(67), 3–11.

Zoet, L.K. and Iverson, N.R. (2020). A slip law for glaciers on deformable beds. *Science*, 368(6486), 76–78.

PART 2

Inversion

4

Probabilistic Inversion Methods

Alexandrine GESRET
Centre de Géosciences, Mines Paris – PSL, Fontainebleau, France

4.1. Local methods versus global methods

Independently of the field of application, the solution of an inverse problem involves the determination of the values of the parameters that characterize a system from experimental measurements. In geophysics, the Earth's properties are deduced from surface observations, such as density distribution from gravimetric measurements, electrical conductivity distribution from apparent resistivity measurements, subsurface wave propagation velocity from wave arrival times measured by seismometers, fault coupling from GPS measurements, etc. This can be reduced to a numerical optimization problem in which the model parameters (characterizing the system) that minimize a cost function (also called misfit function) between measured data and synthetic data are to be estimated. Synthetic data are generated by solving the forward problem that predicts the measurements for a given model. This problem is most often solved using local iterative linearization approaches, which are based on computing the gradient of the cost function. However, these local approaches may not be robust in cases where this cost function is not a smooth function with a well-defined global minimum. It may be multimodal, with a global minimum representing the most probable solution and other minima representing other possible solutions. It may also be more easily affected by some model parameters than by others, presenting valleys for which the cost function varies very little. Linearization assumes a unique solution. To reduce the space of possible solutions, a regularization procedure is applied (Menke 2012). This is

Inversion and Data Assimilation in Remote Sensing,
coordinated by Yajing YAN. © ISTE Ltd 2024.

usually achieved by imposing the solution to be close to an initial a priori model. If the initial solution is too far from the global solution, the inversion can be stopped in a local minimum or in an area where the cost function does not significantly change, as illustrated here for a simplistic example of seismic tomography (Figure 4.1).

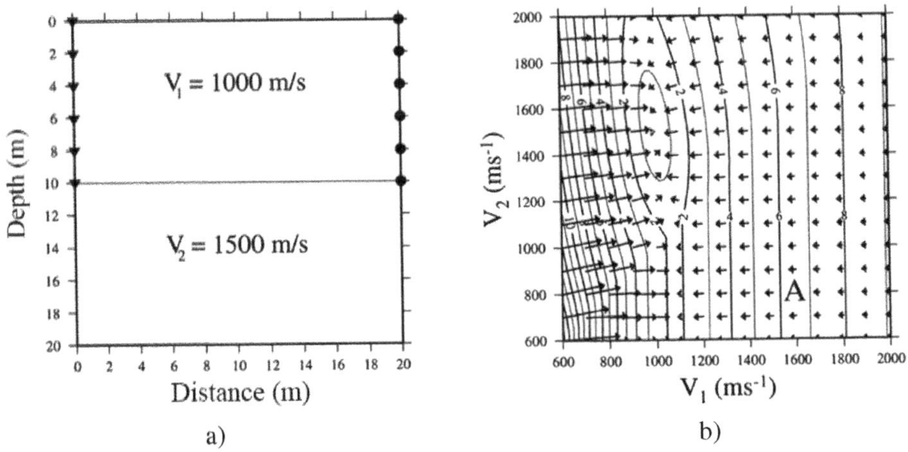

Figure 4.1. *Simple example of seismic tomography. (a) Acquisition geometry and true velocity model. (b) Isovalues of the cost function and descent directions according to the two parameters to be inferred (from Wéber (2000))*

Travel time tomography is a conventional geophysics inverse problem that aims to estimate the subsurface wave propagation velocity model from travel times recorded by an array of seismometers. For this example, the true wave propagation velocity model consists of a simple two-layer model, with six sources set at $X = 0$ m generating waves that are recorded by six sensors deployed at $X = 20$ m (Figure 4.1(a)). Since the depth of the interface is known, the only two parameters to be estimated are the wave propagation velocity in each of the layers. In Figure 4.1(b), the lines represent the values of the cost function and the arrows, the descent directions. Clearly, the solution estimated by a local optimization approach based on the computation of the gradient of the cost function will depend on the initial model. If the initial model is chosen at A, a local approach will converge to an erroneous final solution for which $V_2 = 800$ m/s. Two conclusions can be derived from this very simplistic example: on the one hand, the advantage in using optimization approaches that are not sensitive to the initial model, that is, global optimization approaches. On the other hand, there is an interest in estimating the uncertainties on the parameters to be inverted. For this small example, it would be desirable to find a low uncertainty on parameter V_1 compared to V_2.

Local approaches are not suitable for estimating reliable uncertainties. Rawlinson et al. (2014) give a comprehensive description of existing methods for estimating uncertainties and illustrate that a posteriori probabilistic distribution sampling of model parameters within a Bayesian framework (formalism described in section 4.2) is much more suitable. As a matter of fact, importance sampling enables the cost function to be sampled proportionally to the probability (Tarantola 2005). This sampling is generally carried out using Markov chain Monte Carlo-based algorithms (Mosegaard and Tarantola 1995; Sambridge and Mosegaard 2002), several of which are detailed in section 4.4. The disadvantage of such sampling is the computational cost required by the large number of simulations of the forward problem; thereby, efficient modeling algorithms are needed when considering probabilistic inversion. The number of simulations depends on the number of model parameters to be inverted; therefore, parsimonious parameterizations must be proposed. Section 4.3 introduces some conventional and more original parameterizations.

4.2. Bayesian formalism

Solving the inverse problem is tantamount to a numerical optimization problem in which the model parameters are to be estimated. These latter are contained in the vector $\mathbf{m}$, and minimize a cost function $S(\mathbf{m})$. This function measures the difference between the observed data $\mathbf{d}^{\text{obs}}$ and synthetic data generated for the model $\mathbf{m}$ with the forward problem operator g. The equation relating the model parameters with the observed data is written as:

$$\mathbf{d}^{\text{obs}} = g(\mathbf{m}) + \epsilon \qquad [4.1]$$

where ϵ is called noise and represents all errors, including measurement and modeling errors; in effect, the mathematical model (and also very often the model parameterization) is an approximation of reality.

In most cases, the operator g is nonlinear and non-bijective; therefore, the inverse problem is ill-posed. An inverse problem is ill-posed if it satisfies one of the following conditions: non-existence, non-uniqueness or instability of the solution. In order to solve it optimally, the formulation must take into account all the available a priori information on the parameters to be inferred. It must also represent uncertainties of the observations. To meet these requirements, the probabilistic formulation, in which all information is represented by probability distributions, is the most appropriate. It consists of constructing a distribution called a posteriori distribution representing the probability distribution of the unknown parameters $\mathbf{m} = (m_i)$ conditional on the observed data $\mathbf{d}^{\text{obs}}$. This distribution combines the information contained in the observations and the a priori knowledge about the parameters. The Bayesian framework allows the a priori probability distribution $P(\mathbf{m})$ to be transformed into an a posteriori probability distribution $P(\mathbf{m}|\mathbf{d}^{\text{obs}})$:

$$P(\mathbf{m}|\mathbf{d}^{\text{obs}}) = \frac{P(\mathbf{d}^{\text{obs}}|\mathbf{m})P(\mathbf{m})}{P(\mathbf{d}^{\text{obs}})} \qquad [4.2]$$

The term $P(\mathbf{d}^{\text{obs}})$ that appears in the denominator is called marginal likelihood (or evidence) and represents the probability distribution of the data. This distribution is obtained by integrating the joint distribution of the data with respect to all possible models:

$$P(\mathbf{d}^{\text{obs}}) = \int P(\mathbf{d}^{\text{obs}}, \mathbf{m}).\mathbf{dm} = \int P(\mathbf{d}^{\text{obs}}|\mathbf{m})P(\mathbf{m})d\mathbf{m} \qquad [4.3]$$

This can be considered a normalization constant, since it does not depend on the model (Tarantola 2005).

The *a posteriori distribution* $P(\mathbf{m}|\mathbf{d}^{\text{obs}})$ is therefore known up to a constant and can be written as:

$$P(\mathbf{m}|\mathbf{d}^{\text{obs}}) \propto P(\mathbf{d}^{\text{obs}}|\mathbf{m})P(\mathbf{m}) \qquad [4.4]$$

This distribution is proportional to the product of the a priori distribution $P(\mathbf{m})$ by the likelihood $P(\mathbf{d}^{\text{obs}}|\mathbf{m})$ and summarizes how the information provided by the observed data improves knowledge of the model parameters.

Likelihood $P(\mathbf{d}^{\text{obs}}|\mathbf{m})$ represents the probability of observing the data $\mathbf{d}^{\text{obs}}$ knowing a given model $\mathbf{m}$. This probability is based on the forward problem linking the unknown variables to the data. It represents how the predicted data $g(\mathbf{m})$ fits to the observations and depends on the noise model. A common form of likelihood is the Gaussian likelihood (Mosegaard and Tarantola 1995), for which it is assumed that measurement and modeling errors follow a Gaussian distribution:

$$P(\mathbf{d}^{\text{obs}}|\mathbf{m}) = \frac{1}{\sqrt{(2\pi)^{N_d}|\Sigma_d|}} \exp[-\frac{1}{2}(\mathbf{d}^{\text{obs}} - g(\mathbf{m}))^t \Sigma_d^{-1}(\mathbf{d}^{\text{obs}} - g(\mathbf{m}))] \qquad [4.5]$$

where $g(\mathbf{m})$ represents the predicted data, $|\Sigma_d|$ the determinant of the data covariance matrix. The error variances are therefore contained in the diagonal of this matrix Σ_d. Very often, the covariance matrix is diagonal under the assumption of error independence, and the same noise value is considered for all data. In this case, the formula [4.5] is simplified as:

$$P(\mathbf{d}^{\mathrm{obs}}|\mathbf{m}) = \frac{1}{\sqrt{(2\pi)^{N_d}\sigma_d^{N_d}}} \exp\left[-\frac{1}{2}\left\|\frac{\mathbf{d}^{\mathrm{obs}} - g(\mathbf{m})}{\sigma_d}\right\|^2\right] \quad [4.6]$$

It is important to stress that the Gaussian form of the likelihood does not imply a Gaussian form of the a posteriori distribution. The Gaussian model based on a quadratic norm of the cost function is the most widespread, but this L_2 norm is not robust when addressing data sets containing outliers. Other distributions may be more appropriate in certain cases, such as the Laplace distribution with the L_1 norm, which will perform better when there is a moderate number of outliers (Kijko 1994), for example, for representing the picking error on seismograms (Zhang et al. 2014). The choice of likelihood will have a significant impact on the results, mainly depending on data quality.

The a posteriori distribution strongly depends on the standard deviation of the data σ_d. If its value is uncertain, it may therefore make sense that it be considered as an unknown parameter (hyper-parameter) of the inverse problem. The *hierarchical* Bayesian model introduced in geophysics by Malinverno and Briggs (2004) and Malinverno and Parker (2006) makes it possible to address this problem. One or more additional parameters $\mathbf{h}$ are added to the model parameters, which will be perturbed in the same way as the other model parameters (see section 4.4 on sampling algorithms), with the aim of estimating $P(\mathbf{m}, \mathbf{h}|\mathbf{d}^{\mathrm{obs}})$. The simplest approach is to simply infer the value of σ_d, but other, more elaborate formulations have been proposed, for example, for estimating noise level and correlation between the data (Malinverno and Briggs 2004; Agostinetti and Malinverno 2010; Bodin et al. 2012a).

The a priori distribution $P(\mathbf{m})$ is generally chosen to faithfully transcribe the knowledge about the model available before collecting the observations. This information may originate, for example, from geological knowledge of the environment, or from independent experiments. In practice, it is chosen according to the problem to be solved and depends on the parameterization. The normal distribution is, for instance, often employed, and in this case, the a priori knowledge will consist of an order of magnitude on the value of the parameters associated with an uncertainty interval:

$$P(\mathbf{m}) = \frac{1}{\sqrt{(2\pi)^{N_p}|\Sigma_p|}} \exp[-\frac{1}{2}(\mathbf{m} - \mathbf{m}_{prior})^t \Sigma_p^{-1}(\mathbf{m} - \mathbf{m}_{prior})] \quad [4.7]$$

where N_p is the number of parameters, $\mathbf{m}_{prior}$ is the parameter vector containing the assumed values and $\Sigma_p$is the associated covariance.

It may occur that no relevant information on the parameters is a priori available and that only realistic bounds on the parameter values can be assumed. In such situations, a non-informative distribution should be used, such that it does not give preference to any parameter value, such as the uniform distribution, which describes the lack of information between a minimum and a maximum parameter value:

$$P(m_i) = \frac{1}{\Delta m_i} \text{if } m_i \in [m_{i_{min}}, m_{i_{max}}] \qquad [4.8]$$

where m_i is the ith component of the parameter vector. All values are thus a priori equiprobable. Very generally, the model parameters are assumed to be a priori independent, hence:

$$P(\mathbf{m}) = \prod_{i=1}^{N_p} P(m_i) \qquad [4.9]$$

Bayes' theorem is therefore a means to update the a priori distribution by integrating the physical theory g (which relates the model parameters to the observations) and the observations $\mathbf{d}^{\text{obs}}$ with their associated uncertainties by summarizing the a posteriori distribution as (under the Gaussian distribution assumption):

$$P(\mathbf{m}|\mathbf{d}^{\text{obs}}) \propto P(\mathbf{m}) \exp\left[-\frac{1}{2}(\mathbf{d}^{\text{obs}} - g(\mathbf{m}))^t \Sigma_d^{-1}(\mathbf{d}^{\text{obs}} - g(\mathbf{m}))\right] \qquad [4.10]$$

In this Bayesian framework, either only the most probable model (which maximizes the previous formula) is sought for, which corresponds to *optimization*, or the full shape of the distribution is looked for, which corresponds to *sampling*. Generally, the most probable solution is estimated by a numerical optimization technique: either a local optimization approach based on the computation of the cost function gradient or a global optimization technique. As recalled in section 4.1, the disadvantage of local methods is their strong dependency on the initial model. To overcome this drawback, global optimization methods are preferably used. In the latter case, it is more likely that the solution obtained will be close to the true solution, but the uncertainties will then also have to be estimated based on this solution under certain assumptions, most often under a Gaussian distribution assumption, although the shape of the a posteriori distribution can be far more complex. In order to avoid this assumption, the ideal solution is to estimate the full probability density a posteriori. For a small number of model parameters ($3-4$), the a posteriori probability density function can be estimated through an exhaustive grid search; however, for a larger number of parameters to be inferred, it is necessary to perform importance sampling using, for example, Markov chain Monte Carlo (MCMC) algorithms, which are particularly well-suited to sampling (Robert and Casella 2013). This type of algorithm in fact generates quasi-independent models that are distributed proportionally to the a posteriori distribution (Koren et al. 1991; Mosegaard and Tarantola 1995). In other words, the frequency of occurrence approximates the a posteriori probability density $P(\mathbf{m}|\mathbf{d}^{\text{obs}})$. This sampling of the full distribution thus provides an unbiased estimate of the uncertainties of the velocity model (Debski 2010). The downside is the computation time required by MCMC algorithms. A large number of iterations are generally required before the chain converges to a stable configuration proportional to the a posteriori distribution. In order to reduce the computation time, the procedure generally consists of running several independent chains in parallel and aggregating the results, for instance, Bodin et al. (2012a) ran 96 independent chains for 3 million iterations. Probabilistic approaches therefore require a large number of simulations (generally of the order of a million), and computationally efficient algorithms are thus essential for solving both the forward problem, which will be called many times, and the inverse problem. Section 4.4 presents the classic MCMC-based algorithms, followed by more elaborate algorithms. In addition, since the number of parameters to be inverted should be kept to a minimum, usual and sparse parameterizations are introduced in section 4.3.

4.3. Model parameterization

For earthquake localization, parameterization is straightforward, since the parameters to be inferred are simply the coordinates and time of origin of the earthquake. Similarly, few parameters are needed to describe the mechanism at the earthquake source. Model parameterization becomes an issue in itself when it comes to solving structural imaging inverse problems. Indeed, in probabilistic inversion, it is necessary to use the most sparse parametrization possible to enable sampling, which involves the forward problem a number of times. At the same time, this parameterization must represent reality as accurately as possible. In geophysics, the layered model and Voronoi tessellation are rather widely used. Wavelets and Johnson Mehl tessellation (an extension of the Voronoi tessellation) are also introduced in the following sections. The advantage of layers, or wavelets, is that they form a physical representation of the model, which is not the case with parametric representations produced by tesselations. A single representation has no physical sense: estimators such as the mean and standard deviation must systematically be used. The advantage of averaging over a large number of models is that it enables smoothly varying a posteriori mean models to be obtained. Over the last 10 years, the community has been increasingly utilizing *transdimensional inversion*, where the number of parameters representing the model $\mathbf{m}$ is itself also to be inferred. Transdimensional inversion is naturally attractive, since it gives preference to sparsity, that is, the minimum number of parameters required to fit observations for a given noise level. One of the first applications of transdimensional inversion in geophysics concerned the inversion of resistivity data for a layered model (Malinverno 2002). Since then, numerous applications have followed in $1D$ (e.g. (Agostinetti and Malinverno 2010; Dettmer et al. 2010, 2012)), $2D$ (e.g. Bodin and Sambridge 2009; Bodin et al. 2012a, 2012b) and even in $3D$ (Agostinetti et al. 2015). The most commonly used parameterizations in transdimensional inversion are layered models and Voronoi tessellation. Since the number of parameters is unknown, specific sampling algorithms have to be used, such as the reversible jump Metropolis-Hastings algorithm rj-MCMC (section 4.4.4).

4.3.1. *Layered models*

The layered model is ideally suited to the characterization of layered reservoirs, in surface wave tomography or in receiver function analysis, which prove sensitive at interfaces. In this case, the structure is simply characterized by the depth of the layers, and eventually by their dip angles, and finally their properties, such as wave propagation velocity or resistivity (Figure 4.2).

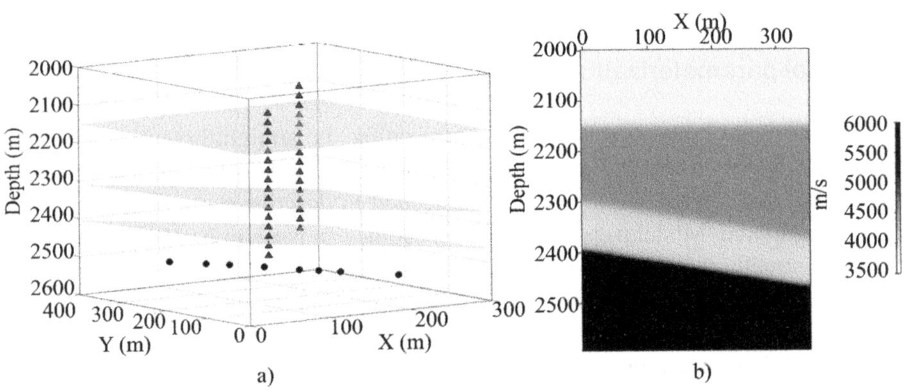

Figure 4.2. *(a) Example of four-layer model. (b) Property in each of the four layers (here, wave propagation velocity) (from Gesret et al. (2015)). For a color version of this figure, see www.iste.co.uk/yan/remotesensing.zip*

The parameter vector to be inferred is then written as: $\mathbf{m} = (\mathbf{z}, \boldsymbol{\alpha}, \boldsymbol{\beta}, \mathbf{v})$, where $\mathbf{z}$ is the vector containing the depths, $\boldsymbol{\alpha}$ and $\boldsymbol{\beta}$ are the vectors containing the dip angles and $\mathbf{v}$ is the vector containing the property within the layer. This parameterization can also be applied in the transdimensional case, where the number of layers is a parameter to be inferred.

4.3.2. *Wavelets*

Wavelet analysis is a well-known tool for multiscale signal analysis, for example, in image compression. Since the 2000s, wavelets have been used to parameterize the model (for instance, Chiao and Kuo 2001; Loris et al. 2007; Tikhotsky and Achauer 2008). Its strong point lies in decomposing a complex signal onto a family of wavelets obtained by dilating and translating a mother wavelet (Mallat and Hwang 1992). As a matter of fact, a function $h(x_j)$ can be approximated by an expansion with N coefficients a_i on a wavelet basis:

$$h(x_j) = \sum_{i=1}^{N} a_i \phi_i(x_j) \quad [4.11]$$

The first coefficients will represent the overall structure (long wavelengths) of function h , while the smallest coefficients represent small-scale variations (short wavelengths). This series is generally truncated, keeping only the most energetic coefficients. There are many different types of wavelets (Haar, Ricker, Daubechies, etc.). The following examples illustrate the decomposition of a sonic log (Figure 4.3) on a wavelet basis (Bottero et al. 2016).

If this sonic log is to be used as a priori information in a probabilistic inversion for the purpose of estimating the velocity profile, this log has necessarily to be parameterized. This parameterization with a layered model is not an easy task, because it may require a large number N of layers (difficult to be determined) and therefore too many parameters to be inverted ($2N - 1$ parameters: velocity to be inferred in each layer and interface depth).

In addition, this log has a very high frequency; other analyses using seismic waves will not have the frequency content needed to solve all these heterogeneities. A sparse parameterization based on a wavelet decomposition of this sonic log may be preferable. As such, the log $V(x_j)$ can be written as:

$$V_M(x_j) = \sum_{i=1}^{M} b_i \phi_i(x_j) \quad [4.12]$$

where b_i are the coefficients and $\phi_i(x_j)$ are the wavelets obtained by dilating and translating an analyzing wavelet. This decomposition makes it possible to use the sonic log as a priori information and to describe this complex velocity profile with a small number of parameters. In fact, with only 15 coefficients, very satisfactory profiles are obtained (red curves in Figure 4.3(c)), especially when decomposing the log on a Daubechie wavelet basis. The advantage of this decomposition is that it reduces the number of parameters to be inverted, since only these 15 coefficients will then need to be inferred, enabling the use of a probabilistic approach.

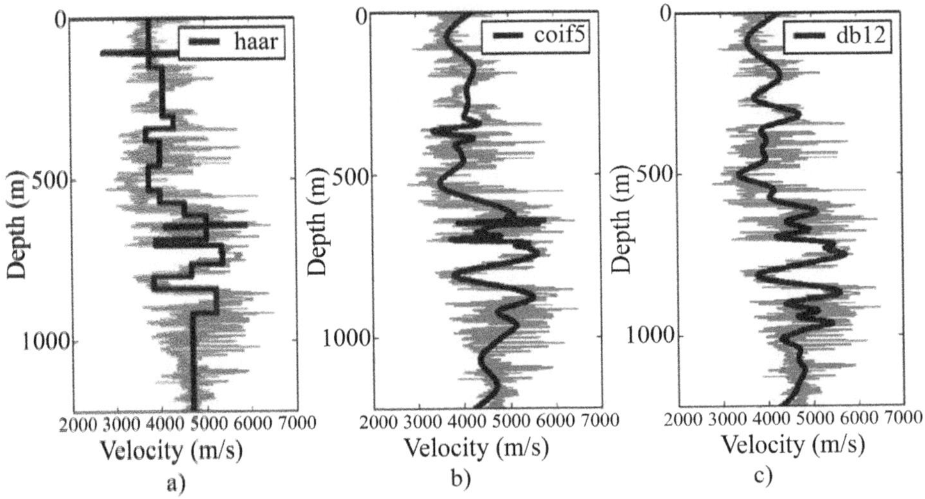

Figure 4.3. *Decomposition of a sonic log (blue) on a family of Haar wavelets (a), Coiflet wavelets (b) and Daubechie wavelets (c) using 15 coefficients (from Bottero et al. (2016)). For a color version of this figure, see www.iste.co.uk/yan/remotesensing.zip*

Hawkins and Sambridge (2015) proposed a multiscale approach using an increasingly detailed decomposition of the domain based on a tree structure. The Markov chain Monte Carlo algorithm (see section 4.4) accepts or rejects the finer-scale decomposition and determines the value of coefficients associated with functions such as wavelets. This transdimensional approach, applicable in $1D$, $2D$ and $3D$, gives preference to a sparse model representation.

Applications of this tree-structure-based transdimensional approach have been proposed for quantifying uncertainties in sea surface displacement in the case of tsunamis (Dettmer et al. 2016), for inversion of airborne electromagnetic data (Hawkins et al. 2018) and for full waveform inversion (Ray et al. 2018). This parameterization, applicable in $2D$ or $3D$, is to-date less widely used than space partition into a Voronoi tessellation.

4.3.3. *Voronoi tessellation*

The Voronoi tessellation is a spatial partition into n cells of kernel c_i; each cell contains the set of points closer to its kernel than to any other kernel (Figure 4.4).

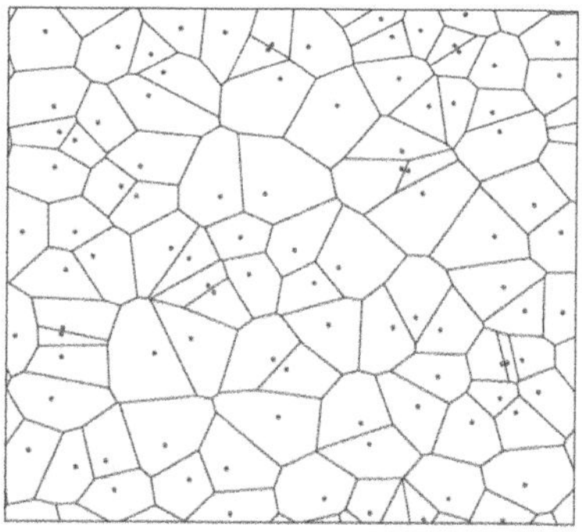

Figure 4.4. *Voronoi tessellation. For a color version of this figure, see www.iste.co.uk/yan/remotesensing.zip*

Each cell is characterized by the coordinates of its core and the value of the parameter sought for in the cell (e.g. resistivity, seismic velocity and density). If the number of cores is fixed, and only the position of the cores and the property value in each cell are variable, then the parameter vector $\mathbf{m}$ to be estimated is expressed as $\mathbf{m} = \mathbf{c}, \mathbf{v}$, where $\mathbf{v}$ represents the vector containing the medium property in each cell. If the number of nuclei is unknown and variable (transdimensional inversion), the vector $\mathbf{m}$ becomes $\mathbf{m} = n, \mathbf{c}, \mathbf{v}$ with n the number of nuclei, the new parameter to be inferred.

The a priori distribution in the case of a transdimensional inversion is decomposed as:

$$P(\mathbf{m}) = P(\mathbf{m}|n)P(n) \quad [4.13]$$

where $P(n)$ is the prior over the number of parameters (cells, in this case). If a uniform distribution is used:

$$P(n) = \frac{1}{\Delta n}\text{if } n \in [n_{min}, n_{max}] \quad [4.14]$$

where $\Delta n = (n_{max} - n_{min})$.

For a given number of nuclei, the nuclei positions (contained in the vector $\mathbf{c}$) and the medium property to be inferred, here we consider the wave propagation velocity $\mathbf{v}$, are independent, thus:

$$P(\mathbf{m}|n) = P(\mathbf{c}|n)P(\mathbf{v}|n) \tag{4.15}$$

A uniform prior can be chosen for the property (equation [4.8]) and if the medium property between each cell is independent:

$$P(\mathbf{v}|n) = \prod_{i=1}^{n} P(v_i|n) \tag{4.16}$$

Also, for example, as proposed by Bodin and Sambridge (2009), the prior for the positions for n cells can be written as:

$$P(\mathbf{c}|n) = \left(\frac{N!}{n!(N-n)!}\right)^{-1} \tag{4.17}$$

The a posteriori distribution to be inferred is written as:

$$P(\mathbf{m}, n|\mathbf{d}^{\text{obs}}) \propto P(\mathbf{d}^{\text{obs}}|\mathbf{m}, n)P(\mathbf{m}|n)P(n) \tag{4.18}$$

4.3.4. *The Johnson Mehl tessellation*

A new parameterization based on Johson Mehl cells has recently been proposed (Belhadj et al. 2018) for application in geophysics. This tessellation was introduced by Johnson and Mehl (1939) to describe the crystallization process. It can be seen as a generalization of the Voronoi tesselation, for which the number of nuclei increases with time.

A point in the domain belongs to the ith cell if and only if:

$$\forall j = 1, .., n/j \neq i, ||\mathbf{u} - \mathbf{u_i}|| + t_i < ||\mathbf{u} - \mathbf{u_j}|| + t_j$$

When the implantation times t_i, $i = 1, ..., n$ are equal to zero, the Voronoi tessellation is recovered. When a nucleus is implanted, the associated cell grows isotropically with a constant velocity; its growth is halted when it encounters another cell. Figure 4.5 illustrates the construction stages of this tesselation. Unlike the Voronoi tessellation, this parameterization contains cells that do not necessarily have

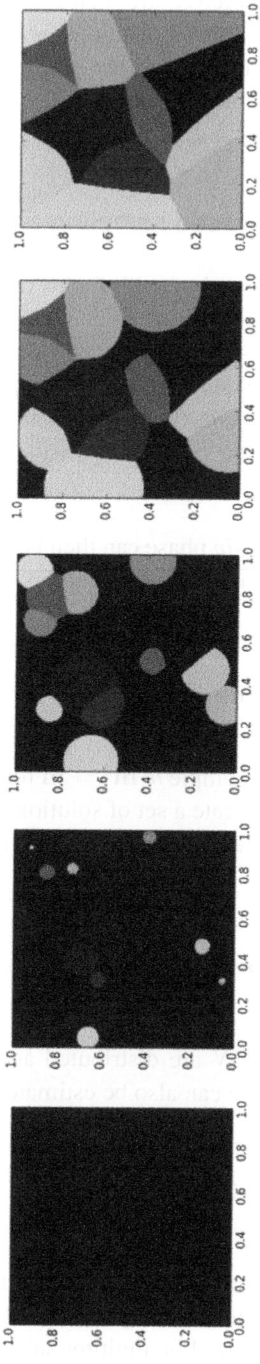

Figure 4.5. *Construction of the Johson Mehl tessellation over time (from Belhadj et al. (2018)). For a color version of this figure, see www.iste.co.uk/yan/remotesensing.zip*

straight sides. The parameters to be inferred are then $\mathbf{m} = n, \mathbf{c}, \mathbf{v}, \mathbf{t}$, where n is the number of cells, the vector $\mathbf{c}$ contains nuclei positions, $\mathbf{v}$ is the vector containing the medium property in each cell and $\mathbf{t}$ is the vector containing the nuclei implementation times. A comparison between the Voronoi and Johnson-Mehl tesselations is illustrated in section 4.4.4.2.

4.4. Markov chain Monte Carlo-based sampling algorithms

As previously mentioned, this section will focus on MCMC-based sampling algorithms. The basic idea is to generate a random sequence of models $\mathbf{m}^0 \rightarrow \mathbf{m}^1 \rightarrow \mathbf{m}^2 \rightarrow \ldots \rightarrow \mathbf{m}^j \rightarrow \ldots \rightarrow \mathbf{m}^n$ whose distribution approximates $P(\mathbf{m}|\mathbf{d}^{\text{obs}})$, each new state of the chain $\mathbf{m}^{j+1}$ depending only on the current state $\mathbf{m}^j$. Before convergence to the a posteriori distribution, the chain is in a *burn-in* phase, during which the probability (respectively, the cost function) gradually increases (respectively, decreases).

The samples visited after the *burn-in* phase can then be used to calculate statistical a posteriori estimators of any h function. These quantities of interest are described by integrals of the form:

$$E[h(\mathbf{m}|\mathbf{d}^{\text{obs}})] = \int h(\mathbf{m})P(\mathbf{m}|\mathbf{d}^{\text{obs}})dm \qquad [4.19]$$

where h is a function of interest, for example $h(\mathbf{m}) = \mathbf{m}$ for the a posteriori mean. The aim of MCMC methods being to generate a set of solutions distributed proportionally to the a posteriori distribution, the integral (equation [4.19]) can then be approximated by:

$$E[h(\mathbf{m}|\mathbf{d}^{\text{obs}})] = \frac{1}{N}\sum_{j=1}^{N} h(\mathbf{m}^j) \qquad [4.20]$$

where the samples $\mathbf{m}^j$, $j = 1, 2, \ldots N$ are distributed according to the a posteriori distribution $P(\mathbf{m}|\mathbf{d}^{\text{obs}})$. The variance can also be estimated as:

$$Var[h(\mathbf{m}|\mathbf{d}^{\text{obs}})] = \frac{1}{N}\sum_{j=1}^{N}(h(\mathbf{m}^j) - E[h(\mathbf{m}|\mathbf{d}^{\text{obs}})])^2 \qquad [4.21]$$

In some cases, it may make more sense to estimate the maximum a posteriori rather than the mean and standard deviation, given that the a posteriori distribution generally exhibits no Gaussian shape. For multimodal solutions, the mean is not relevant. However, it may prove more reliable than the maximum a posteriori, since the summation of several models results in filtering out the random part of the inversion errors and in only retaining the characteristics common to all models (Debski 2010).

Marginal a posteriori distributions are much more informative than the estimators previously mentioned. A single-parameter $1D$ marginal distribution is defined as the integral over all other parameters:

$$P(m_i|\mathbf{d}^{\text{obs}}) = \int P(\mathbf{m}|\mathbf{d}^{\text{obs}}) \prod_{j \neq i} dm_j \qquad [4.22]$$

Similarly defined $2D$ marginal distributions can also be easily used, in particular to observe the correlation between parameters. Marginal distributions can then be used to easily calculate quantiles (e.g. 68% or 95%) by ordering the parameters from maximum probability to minimum probability.

Figure 4.6 represents various ways of representing results, including uncertainties, obtained from probabilistic inversion. An example of travel time tomography is depicted here, where the parameters to be inferred are the parameters (depth, dip and velocity) of a layered model (Figure 4.2) for observed travel time data generated by eight sources (black circles in Figure 4.2(a)) recorded by 32 sensors (red triangles in Figure 4.2(a)). As an example, the best model (maximum probability), quantiles, mean model, the associated standard deviation and marginal distributions are shown.

Regardless of the estimator that is used, aggregation of the results requires that the chain have converged to the a posteriori distribution. Several methods have been proposed for testing chain convergence (such as in Brooks and Roberts 1998; Robert and Casella 2013), the most widely used of which are listed here. The first test is the chain acceptance rate, which in theory should be close to 0.234 for dimensions greater than 5 and for independent and identically distributed variables (Roberts et al. 1997) or simply independent variables (Rosenthal 2009). In practice, an acceptance rate of between 0.25 and 0.5 is generally searched for. A chain must have good *mixing* properties, more specifically the states must not be too correlated. In practice, the auto-correlation function per parameter can be calculated, which should decrease with the number of iterations and tend toward zero for a reasonable number of samples.

One method of avoiding correlation is *thinning*, for example, by keeping only one sample in every 100. To test chain convergence, some use the Gelman–Rubin diagnostic (Gelman and Rubin 1992), which compares the variance within each chain with the variance between several Markov chains, which are supposed to be similar and therefore give a ratio between the two variances close to 1. Gelman and Rubin

(1992) proposed a formula for calculating the potential scale reduction factor (PSRF); a value of this ratio below 1.2 indicates that the chain has converged toward the a posteriori distribution (Brooks and Gelman 1998).

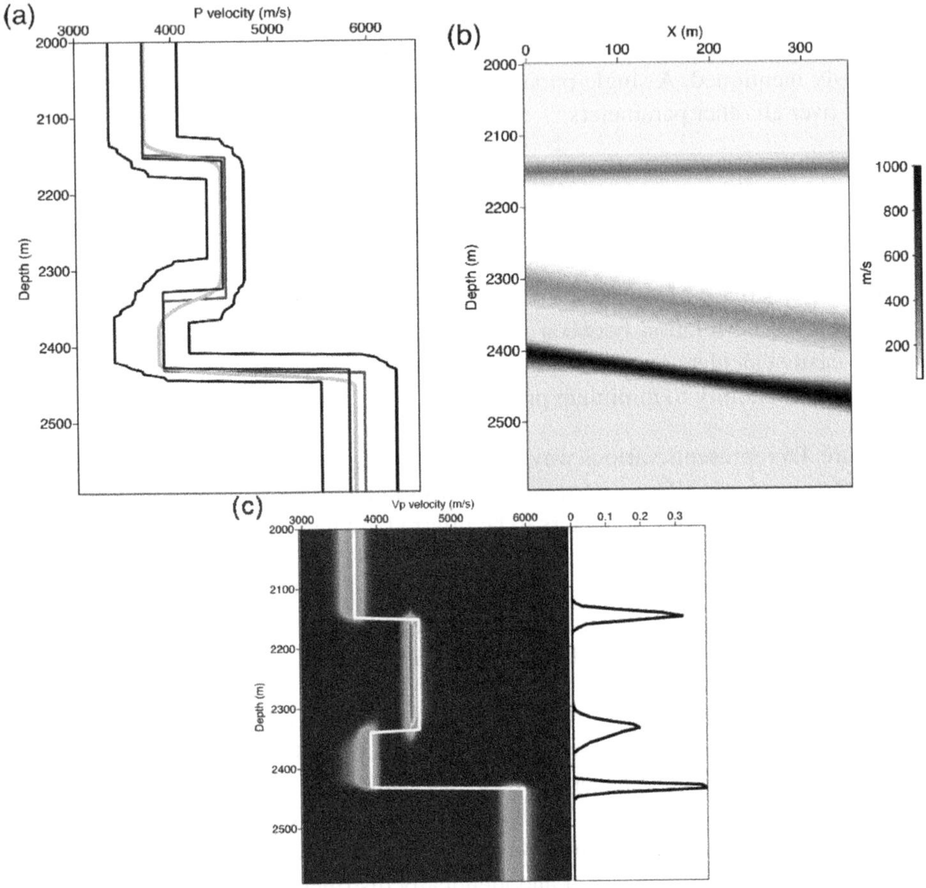

Figure 4.6. *Representations of results including uncertainties for an example of travel time tomography. The inferred parameters are depth, dip and velocity of a layered model. For a color version of this figure, see www.iste.co.uk/yan/remotesensing.zip*

COMMENT ON FIGURE 4.6.– *(a) Vertical profiles of the true model (red), the mean model (green), the best model (blue) and quantiles 95%. (b) Vertical cross-section of the standard deviation associated with the mean model. (c) Density plot of the marginal velocity distribution where warm (respectively, cold) colors correspond to high (respectively, low) probability, and marginal distributions on the depths of the three interfaces are also shown (from Gesret et al. 2015).*

4.4.1. *The Metropolis-Hastings algorithm*

The best-known MCMC algorithm for sampling a probability distribution is the Metroplis-Hastings algorithm (Metropolis et al. 1953; Hastings 1970). Its implementation is simple, involving two steps: a new state is proposed followed by the computation of the acceptance probability.

4.4.1.1. *Proposition distribution*

From a current chain state $\mathbf{m^j}$, a new $\mathbf{m}^*$ state is proposed according to a proposition distribution $q(\mathbf{m}^*|\mathbf{m^j})$ that describes the probability of proposing a state $\mathbf{m}^*$ knowing that the chain is in state $\mathbf{m^j}$. The two best-known distributions are:

– the independent sampler IMH $q(\mathbf{m}^*|\mathbf{m^j}) = q(\mathbf{m}^*)$ where q is generally the a priori distribution $P(\mathbf{m})$ that follows a uniform distribution for which all states of the a priori distribution are equiprobable (equation [4.8]);

– the symmetrical random walk sampler SIMH $q(\mathbf{m}^*|\mathbf{m^j}) = q(|\mathbf{m}^* - \mathbf{m^j}|)$, which very generally follows a normal distribution centered on the value of the current state and with a standard deviation set by the user, $q = N(\mathbf{m^j}, \boldsymbol{\sigma}^2_{\mathbf{m}})$.

4.4.1.2. *Acceptance probability*

The second step consists of calculating the acceptance probability, which depends on the ratio of the probability density of the new state to the probability density of the current state:

$$\alpha(\mathbf{m}^*|\mathbf{m^j}) = min\left(1, \frac{P(\mathbf{m}^*|\mathbf{d^{obs}})q(\mathbf{m^j}|\mathbf{m}^*)}{P(\mathbf{m^j}|\mathbf{d^{obs}})q(\mathbf{m}^*|\mathbf{m^j})}\right) \quad [4.23]$$

For the SIMH or uniform IMH, the formula is simplified as:

$$\alpha(\mathbf{m}^*|\mathbf{m^j}) = min\left(1, \frac{P(\mathbf{m}^*|\mathbf{d^{obs}})}{P(\mathbf{m^j}|\mathbf{d^{obs}})}\right) \quad [4.24]$$

In practice, a random number ρ between 0 and 1 is generated. If $\rho < \alpha(\mathbf{m}^*|\mathbf{m^j})$, the transition is accepted, $\mathbf{m^{j+1}} = \mathbf{m}^*$; otherwise, the chain remains in its current state $\mathbf{m^{j+1}} = \mathbf{m^j}$. If the new state has a higher probability, it will necessarily be accepted. On the other hand, the lower the probability of the new state compared to the current state, the lower the probability of accepting it.

4.4.1.3. *Illustration*

Figure 4.7(a) illustrates the two samplers (IMH and SIMH) for a problem in which the a posteriori distribution to be inferred is bimodal.

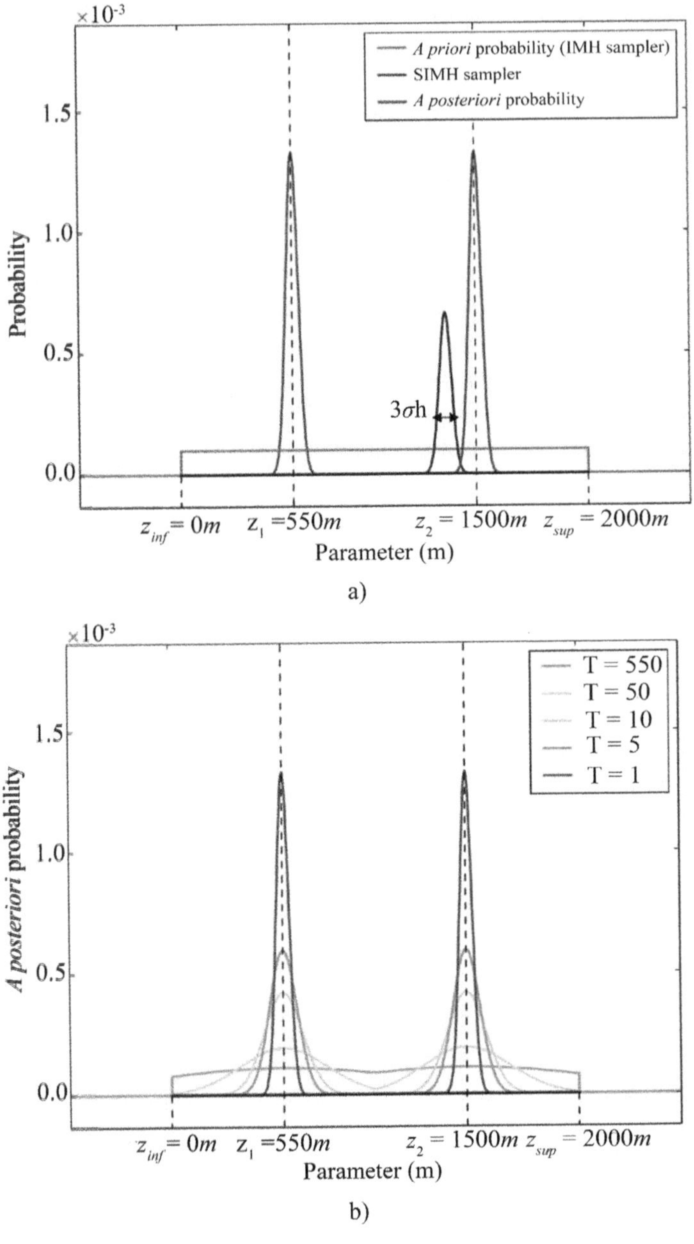

Figure 4.7. *(a) Example of a bimodal distribution (in blue) with two samplers (IMH in red and SIMH in black). (b) Effect of temperature on the tempered a posteriori distribution (from Bottero et al. (2016)). For a color version of this figure, see www.iste.co.uk/yan/ remotesensing.zip*

It appears that with an independent sampler, exploration will certainly be coarse, and it will be difficult to identify modes since the propositions will be independent of the previous state. For the SIMH, the exploration will depend very much on the width of the standard deviation; if it is too small, the chain will have poor mixing properties, because the states will be highly correlated, and almost all states will be accepted. Moreover, convergence will be very slow, and the chain may get stuck in one mode when dealing with a multimodal distribution. If the standard deviation is too large, areas with too low a probability will be tested and too many states will be rejected. The chain may succeed in detecting modes, but they will not be finely sampled.

Adjusting this parameter is therefore not so easy, and in practice it is set by trial-and-error monitoring the chain acceptance rate. One way of overcoming this issue is to use the Metropolis adaptive algorithm (Haario et al. 2001), which results in reducing computation time by updating the covariance of the sampler distribution using information accumulated during previous states.

The samples obtained by the chain after convergence are then used to compute classical statistical estimators. The Metroplis-Hastings algorithm is simple to implement, but its efficiency in terms of computational time is limited. To accelerate convergence, simulated annealing, introduced by Metropolis et al. (1953) and then generalized by Kirkpatrick et al. (1983), may be preferred.

4.4.2. *Simulated annealing*

By adding a parameter, the temperature T, the a posteriori distribution is deformed, which facilitates exploration. The algorithm then aims to explore the tempered distribution $P_T(\mathbf{m}|\mathbf{d}^{\mathrm{obs}})$ by defining the energy such that:

$$E(\mathbf{m}) = \frac{1}{2}\left\|\frac{\mathbf{d}^{\mathrm{obs}} - g(\mathbf{m})}{\sigma_d}\right\|^2 - \ln(P(\mathbf{m})) \tag{4.25}$$

with:

$$P_T(\mathbf{m}|\mathbf{d}^{\mathrm{obs}}) \propto \exp\left[-\frac{E(\mathbf{m})}{T}\right] \tag{4.26}$$

For a symmetrical sampler, the acceptance probability is written as:

$$\alpha(\mathbf{m}^{\mathbf{j}}, \mathbf{m}^*) = min\left(1, \exp\left[\frac{E(\mathbf{m}^{\mathbf{j}}) - E(\mathbf{m}^*)}{T}\right]\right) \tag{4.27}$$

It appears that for high temperatures, $T \to \infty$, $\alpha \to 1$, all states will be accepted; whereas, for low temperatures $T \to 0$, only the most probable models will be.

Figure 4.7(b) illustrates the effect of temperature on a simple bimodal distribution. At high temperatures (red curve), the distribution to be inferred is flattened, and the chain can easily "stroll" within this distribution. In practice, the temperature decreases during iterations to accelerate the *burn-in* phase. In theory, the algorithm will converge toward the global minimum for a fairly high initial temperature and a cooling rate $T \propto log(n)$ with n being the iteration. This rate is not applicable, and geometric decay is generally adopted. The choice of maximum temperature is generally based on the acceptance rate (Robert and Casella 2013).

An example implementation proposed by Wéber (2000) is shown as follows:

1) Choose an initial temperature value as small as possible, but such that all states are accepted.

2) Perturb the model parameter with a sampler. If the new parameter is in the prior, compute the acceptance probability; otherwise, generate a new model. Increase the number of simulation steps L_k to T_k by 1.

3) Compute the probability of acceptance (equation [4.27]).

4) Number of simulation steps L_k to T_k:

a) L_k such that the number of accepted steps $A_k = A_{min}$ (e.g. $A_{min} = 20$);

b) $L_k < L_{max}$ (e.g. $L_{max} = 40$) for low temperature values where the acceptance probability is low.

5) Decrease in temperature when $A_k = A_{min}$ or $L_k = L_{max}$, for example $T_{k+1} = 0.99T_k$.

6) Stopping criterion when the number of accepted steps is zero for a number U_{max} (e.g. $U_{max} = 100$) of consecutive temperatures T_k.

Simulated annealing is conventionally used for optimization purposes, in which the best model is reached when the temperature is close to 0. When the temperature is equal to 1, the tempered distribution is the true a posteriori distribution and the latter can then be sampled with Metropolis-Hastings exploration by simply letting the algorithm run. For a multimodal distribution, if cooling has been too fast and the chain is in a local energy minimum, the probability of moving and sampling around the global minimum is very low, hence the interest of multi-chain interaction approaches.

4.4.3. *Interacting Markov chains*

4.4.3.1. *Method*

One approach that enables escaping a local minimum consists of having several interacting chains, as proposed by Geyer (1991), which also speeds up convergence. The process generates K Markov chains in parallel, each at a different temperature (different energy) and allows them to exchange their configuration, that is, their current or past state. The aim is to enable low-temperature chains, particularly the first chain ($T_1 = 1$) that represents the distribution of interest $P(\mathbf{m}|\mathbf{d}^{\text{obs}})$, to avoid getting stuck in local modes by switching between different modes. High-temperature chains are in fact able to better explore the space (Figure 4.7(b)). Through state exchanges, low-temperature chains can then access distribution states that they would not have been able to reach if these exchanges had not taken place.

Sambridge (2014) proposed a geophysical application of *Parallel Tempering* for which exchanges between simultaneous states of chains at non-adjacent temperatures were proposed. According to the work of Romary (2009), interacting chains based on population importance resampling (PIR) introduced by Andrieu et al. (2011), for which exchanges in the history of the adjacent chain at the higher temperature are allowed, are more efficient. The method detailed in Bottero et al. (2016) is therefore introduced here.

Consider a sequence of temperatures $T_1 = 1 < T_2 < ... < T_k = T_{max}$. The a posteriori distribution tempered at a given temperature T_k is written as:

$$P_k(\mathbf{m}|\mathbf{d}^{\text{obs}}) \propto \mathbf{exp}\left(-\frac{E(\mathbf{m})}{T_k}\right) \qquad [4.28]$$

At each iteration n of the algorithm, an exchange movement between two adjacent chains k and $k+1$ with $k \in [1, K]$ is proposed, according to a certain probability Θ:

1) If the proposition is rejected, a classic Metropolis-Hastings transition is performed.

2) If the proposal is accepted, a movement allowing a past state of the chain with higher temperature is proposed. This step is described below.

A state $\mathbf{m}^{(k+1,j)}$ with $j < n$ is proposed by resampling from all the states generated by the chain $k+1$ in the past $\{\mathbf{m}^{(k+1,1)}, \mathbf{m}^{(k+1,2)}, ..., \mathbf{m}^{(k+1,n-1)}\}$ weighted by the value of the importance function: $\{r(\mathbf{m}^{(k+1,1)}), r(\mathbf{m}^{(k+1,2)}), ..., r(\mathbf{m}^{(k+1,n-1)})\}$. This importance function associated with the $\mathbf{m}^{(k+1,j)}$ is written as (Robert and Casella 2013):

$$r^{(k)}(\mathbf{m}^{(k+1,j)}) = \frac{P_k(\mathbf{m}^{(k+1,j)}|\mathbf{d}^{\text{obs}})}{P_{k+1}(\mathbf{m}^{(k+1,j)}|\mathbf{d}^{\text{obs}})} \quad [4.29]$$

$$= \mathbf{exp}\left(-E(\mathbf{m}^{(k+1,j)})\left(\frac{1}{T_k} - \frac{1}{T_{k+1}}\right)\right) \quad [4.30]$$

The exchange is then accepted with probability β:

$$\beta(\mathbf{m}^{(k,n)}, \mathbf{m}^{(k+1,j)}) = \mathbf{min}\left(1, \frac{r^{(k)}(\mathbf{m}^{(k+1,j)})}{r^{(k)}\mathbf{m}^{(k,n)}}\right) \quad [4.31]$$

$$= \mathbf{min}\left(1, \exp\left(\left(\frac{1}{T_k} - \frac{1}{T_{k+1}}\right)(E(\mathbf{m}^{(k,n)}) - E(\mathbf{m}^{(k+1,j)}))\right)\right) \quad [4.32]$$

If the energy (or probability) of the proposed state $\mathbf{m}^{(k+1,j)}$ is lower (or higher) than that of the state $\mathbf{m}^{(k,n)}$, the change of state will therefore be systematically accepted.

The efficiency of the algorithm depends on the type of sampler used for each chain during conventional Metropolis-Hastings transitions. It is indeed obvious that the higher the temperature, the easier it is for the associated sampler to move, because the target distribution becomes flattened under the effect of temperature (Figure 4.7(b)) and, as such, easy to walk through. A global sampler (such as IMH) is therefore the most suitable, as it enables the space to be easily explored. For low-temperature chains, a local sampler (SIMH, for instance) can interestingly be used. As a matter of fact, this local exploration makes it possible to sample more precisely the relevant regions discovered by high-temperature chains during previous iterations. The standard deviations of SIMH samplers can also decrease with temperature. As in all inference approaches based on MCMC algorithms, parameters need to be set; here these are: the number of chains, the maximum temperature, the sequence of temperatures, the sampler at each temperature and the probability of proposing an exchange. The algorithm is therefore not extremely easy to use, but certain rules can be followed (Romary 2010). For example, the number of chains must be sufficient to allow overlapping of the temperate energies. For the choice of the maximum temperature, a single chain with temperature equal to 1 sampling the prior with a uniform distribution can be run for around 500 iterations. The resulting energy distribution is then used to select T_{max} so as to obtain an acceptance rate between 0.6 and 0.75 for this highest temperature. A geometric sequence is then generally chosen for the temperature sequence.

Regarding estimators, in Kou et al. (2006), it is recommended to use the weighted average of all candidates generated by all K chains to estimate the mean and variance:

$$E[h(\mathbf{m}|\mathbf{d}^{\text{obs}})] = \frac{1}{K}\sum_{k=1}^{K} \frac{\sum_{j=1}^{N} \eta^{k,j} h(\mathbf{m}^{k,j})}{\sum_{j=1}^{N} \eta^{k,j}} \quad [4.33]$$

where:

$$\eta^{k,j} = \exp\left(-E(\mathbf{m}^{k,j})\left(1 - \frac{1}{T_k}\right)\right) \quad [4.34]$$

4.4.3.2. *Illustration*

The interacting chains algorithm is compared with a classic simulated annealing run (Bottero et al. 2016) for a multimodal distribution (Figure 4.9(a)). In this example, three interacting chains are run for 10,000 iterations, while the annealing has run for 30,000 iterations. Energies overlapping between the three chains can be observed (Figure 4.8(a)), ensuring a satisfactory exchange rate between the chains. The convergence speed of the interacting chains algorithm is much greater than that of simulated annealing, as shown by the evolution of energy over the iterations (Figure 4.8(b)). More than 1,000 iterations are needed for simulated annealing to converge; whereas, no burn-in phase is required for interacting chains.

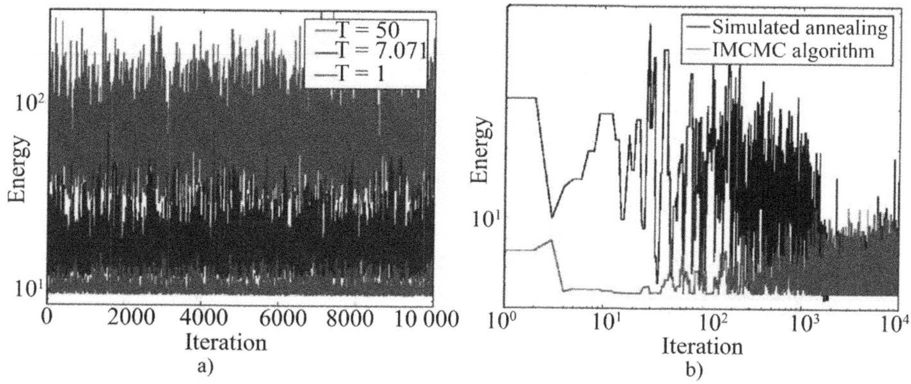

Figure 4.8. *(a) Energies of the interacting chains according to the number of iterations. (b) Convergence in terms of energy for the lowest temperature chain of the interacting chains (red) and the simulated annealing (black) (from Bottero et al. (2016)). For a color version of this figure, see www.iste.co.uk/yan/remotesensing.zip*

For this example of a complex distribution, simulated annealing results in an a posteriori distribution (Figure 4.9(b)) that does not identify all the modes, and the global mode does not correspond to the principal mode. The interacting chain algorithm outperforms simulated annealing since it leads to identifying all modes, and the global mode as the principal mode (Figure 4.9(c)).

This algorithm is therefore much more efficient, stable and robust for estimating uncertainties.

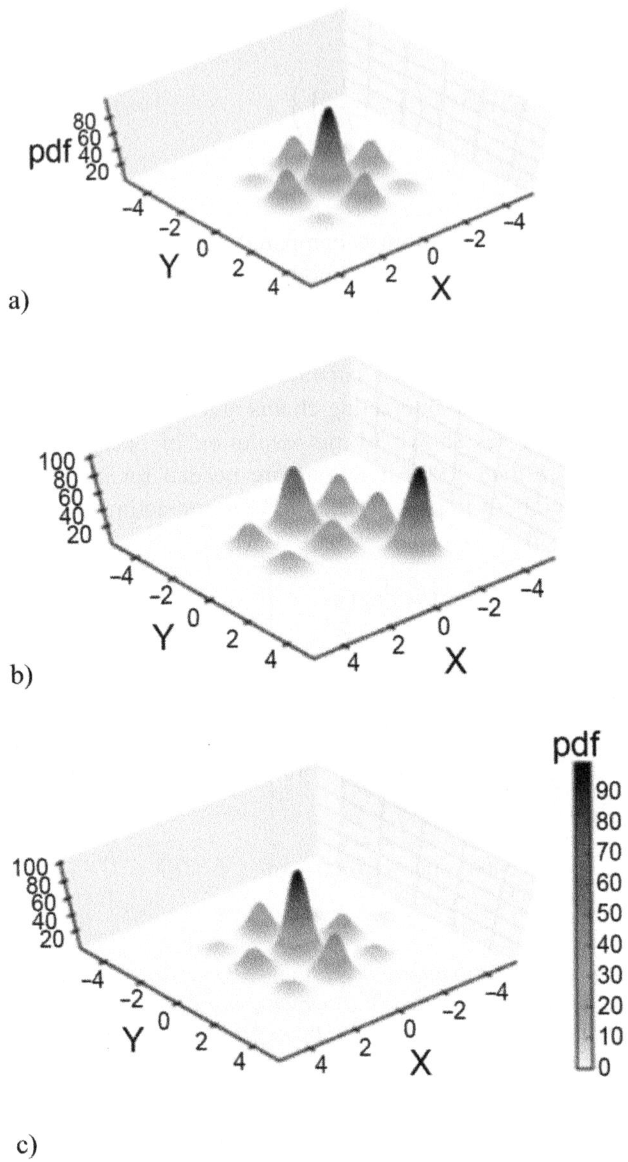

Figure 4.9. *(a) Distribution to be inferred. (b) A posteriori distribution obtained by simulated annealing. (c) A posteriori distribution obtained with the IMCMC algorithm (from Bottero et al. (2016)). For a color version of this figure, see www.iste.co.uk/yan/remotesensing.zip*

4.4.4. *The reversible jump Metropolis-Hastings algorithm*

4.4.4.1. *Method*

The reversible-jump MCMC algorithm is an extension of the Metropolis-Hastings algorithm, which takes into account the variable dimension of the parameter space introduced by Green (1995). The major advantage is that the complexity of the model is inferred by the observed data. The modification that allows transdimensionality to be taken into account is found in the acceptance probability formula (Green 1995; Gallagher et al. 2009). The chain will indeed converge to the transdimensional a posteriori distribution $P(\mathbf{m}|\mathbf{d}^{\text{obs}})$ if:

$$\alpha(\mathbf{m}^*|\mathbf{m}^j) = \min\left(1, \frac{P(\mathbf{m}^*|\mathbf{d}^{\text{obs}})q(\mathbf{m}^j|\mathbf{m}^*)}{P(\mathbf{m}^j|\mathbf{d}^{\text{obs}})q(\mathbf{m}^*|\mathbf{m}^j)} \times |\mathbf{J}|\right) \qquad [4.35]$$

$$\alpha(\mathbf{m}^*|\mathbf{m}^i) = \min\left(1, \frac{P(\mathbf{d}|\mathbf{m}^*, n^*)P(\mathbf{m}^*|n^*)P(n^*)q(\mathbf{m}^i, n^i|\mathbf{m}^*, n^*)}{P(\mathbf{d}|\mathbf{m}^i, n^i)P(\mathbf{m}^i|n^i)P(n^i)q(\mathbf{m}^*, n^*|\mathbf{m}^i, n^i)} \times |\mathbf{J}|\right) \quad [4.36]$$

where $\mathbf{J}$ is the Jacobian of the transformation from $\mathbf{m}^j$ to $\mathbf{m}^*$ which is required to take into account the change of scale when the transformation involves a change of dimension. Typically, $|\mathbf{J}| = 1$. More details on this algorithm are given in Green (2003) and Denison et al. (2002).

A simplification of an example (Bodin and Sambridge 2009) of the detailed proposition distribution is given here for the Voronoi tesselation, but this can quite easily be applied to a layered model where the number of layers is variable:

1) Birth: propose a new cell whose position is chosen randomly in space. The new property value can be chosen according to a proposed normal distribution centered on the old value occupied at this point (probability $1/6$).

2) Death: remove a cell chosen at random (probability $1/6$).

3) Displacement: perturb the position of a cell according to a normal distribution centered on the current position (probability $1/6$).

4) No tessellation disturbance: only the property value inside the cell is modified using a normal distribution centered on the current value (probability $1/2$).

The standard deviations of the proposition distribution must be adjusted. For example, Bodin et al. (2012b) set them by trial-and-error in order to obtain acceptance rates of 44% for each type of perturbation (Rosenthal 1999).

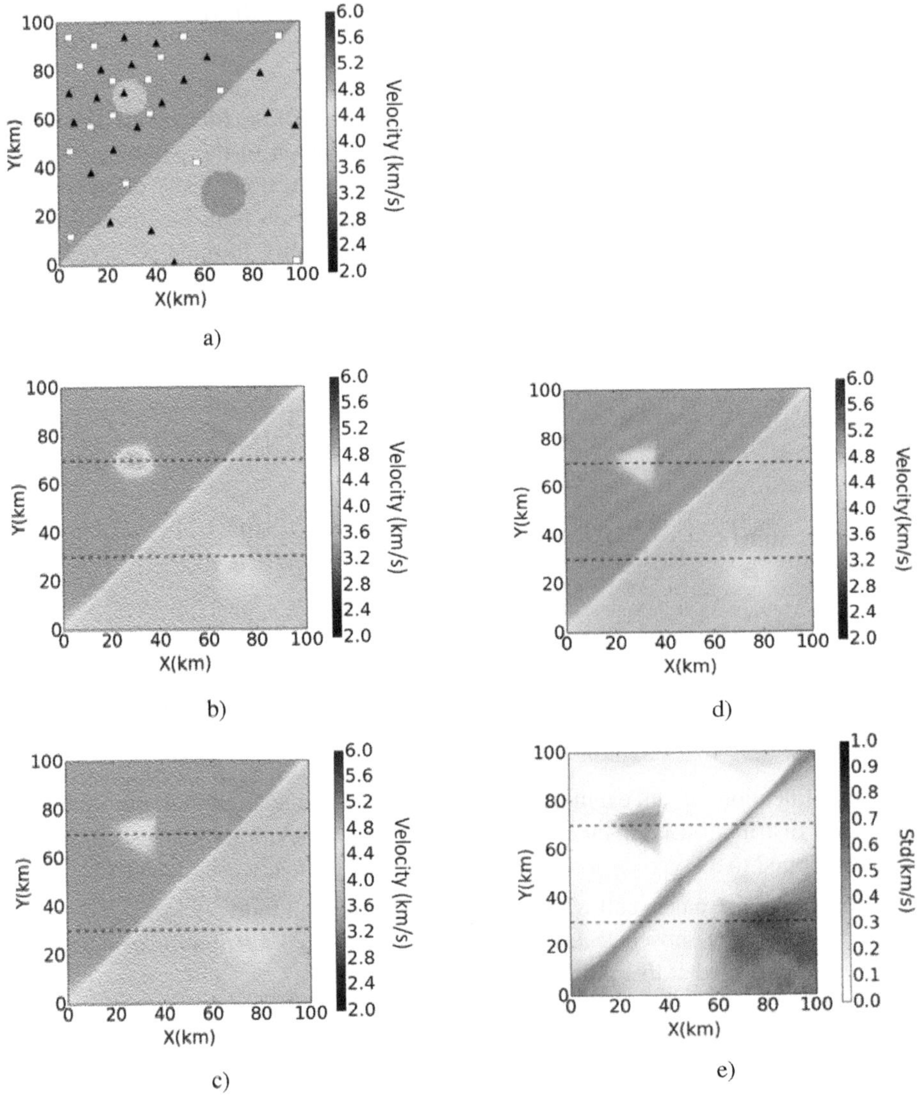

Figure 4.10. *(a) True velocity model and position of the 17 sources (white squares) and 20 receivers (black triangles). Results obtained with Johnson Mehl (b and c) and Voronoi (d and e) parameterization in terms of mean model (b anf d) and standard deviation (c and e) (from Belhadj et al. (2018)). For a color version of this figure, see www.iste.co.uk/yan/remotesensing.zip*

The acceptance probability must then be calculated. The determinant of the Jacobian involved in the acceptance probability (equation [4.36]) is equal to 1 for a situation where the current state $\mathbf{m}^j$ and the proposed state $\mathbf{m}^*$ have the same dimension. For cell birth and death movements, the approach followed by Agostinetti and Malinverno (2010) shows that its determinant is also 1. An example of the reversible jump Metropolis-Hastings algorithm is given below.

4.4.4.2. *Illustration*

One of the first outstanding applications of transdimensional probabilistic inversion in 2D with a reversible jump Metropolis-Hastings algorithm was obtained by Bodin and Sambridge (2009). The authors then used a parameterization of the velocity model based on Voronoi cells. This type of parameterization is now widely used by the scientific community.

A comparison between parameterization with a Voronoi tessellation and a Johnson Mehl tessellation is illustrated here. This test is carried out for an example of seismic tomography, where the property to be inferred is the wave propagation velocity model. The data consist of a set of travel times recorded by receivers for a simple synthetic model (Figure 4.10a), proposed by Bodin and Sambridge (2009).

The wave propagation velocity model is separated into two parts (faster velocity in the upper left than in the lower right) by a diagonal interface, with the upper left (respectively, lower right) containing a low (respectively, high) velocity anomaly. The acquisition geometry consists of 17 sources (white squares) and 20 sensors (black triangles), unevenly distributed. The forward problem of estimating travel times for a given model is solved with an Eikonal solver (Noble et al. 2014).

Transdimensional inversion results are shown for a parameterization based on Johson Mehl cells (Figures 4.10(b) and (c)) and Voronoi cells (Figures 4.10(d) and (e)). These two tesselations perform remarkably very well in faithfully imaging diagonal spatial discontinuity. In this example, the illustrations show an improvement in the quality of the results when the Johnson Mehl tessellation is used, particularly at the level of the slow anomaly. As a result, the shape of the anomaly and the value of its velocity are better found. Moreover, most seismic rays do not cross this anomaly, thereby a high degree of uncertainty is expected in this area, a result effectively obtained with parameterization based on Johnson Mehl cells.

The quality of the results obtained with the two parameterizations is also compared in terms of efficiency and convergence (Figure 4.11).

It appears that the energy decreases faster and reaches a lower level with Johnson Mehl cells, illustrating a significant improvement in efficiency for this example.

Convergence is also estimated using the Gelman-Rubin diagnostic (see section 4.4), which compares the variance within each chain to the variance between several chains that are assumed to be similar. Figure 4.11 illustrates that with the Voronoi parameterization, chain stabilization is not yet achieved, since most values are greater than 1.2; whereas, it is well stabilized with the Johnson Mehl tesselation.

4.5. Conclusions and outlook

The probabilistic approaches that we introduced are attractive because they lead to a set of solutions and therefore reliable uncertainties. Nonetheless, they are currently limited to a few hundred parameters, making it difficult to consider 3D applications. It is important to stress that local methods are highly efficient and can therefore be applied to huge datasets and models containing around a million parameters. Recently, intermediate methods seem to be emerging with the development of new algorithms for inverting a larger number of parameters. The application of the Monte Carlo Hamiltonian to inverse problems in geophysics (Sen and Biswas 2017; Fichtner et al. 2019) certainly represents an interesting approach for extending probabilistic inversion to a much larger number of parameters. However, it still requires the derivatives of the direct problem equations to be computed efficiently and accurately, enabling convergence to be improved and independent samples to be proposed more effectively. In a somewhat similar approach, Zhao and Sen (2020) have very recently inverted up to $20,000$ parameters in a full waveform inversion problem by combining gradient and Hessian computation with a Metropolis-Hastings algorithm. This approach is also attractive, but at the cost of a necessarily more complex implementation.

An essential ingredient of probabilistic inversion is the choice of parameterization, which affects the inferred structure and uncertainties (Hawkins and Sambridge 2015). Hawkins et al. (2019) illustrate, for example, that Voronoi parameterization is more suitable for fields containing discontinuities, while parameterizations based on Delaunay triangulation combined with interpolation would be more suitable for naturally smoother fields such as temperature or density. For probabilistic inversions, the parameterization thus depends on the object we aim to characterize.

One of the great advantages of the probabilistic approach is that it naturally makes it possible to pose the joint inversion of different types of data (as in, for example, Bodin et al. 2012b). If the observations are independent, the a posteriori distribution is simply proportional to the a priori distribution on the parameters to be inferred multiplied by the product of the likelihoods $P(\mathbf{d}_1^{\text{obs}}, \mathbf{d}_2^{\text{obs}}|\mathbf{m}) = P(\mathbf{d}_1^{\text{obs}}|\mathbf{m})P(\mathbf{d}_2^{\text{obs}}|\mathbf{m})$.

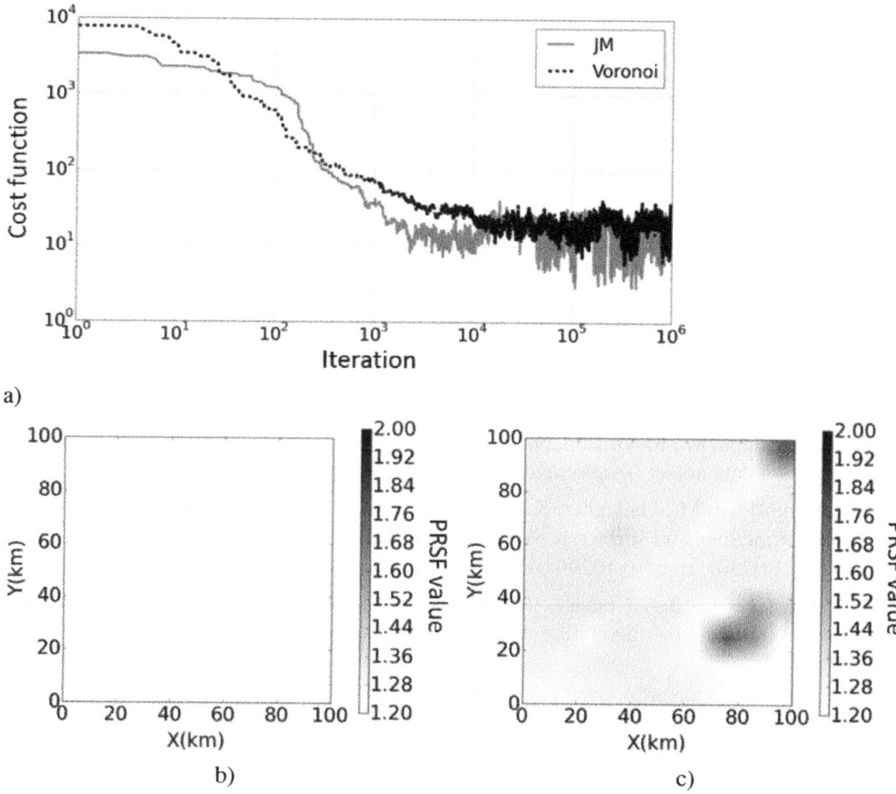

Figure 4.11. *(a) Comparison of the evolution of the cost function according to iterations for a parameterization based on Johnson Mehl (respectively, Voronoi) cells in gray (respectively, black). Gelman-Rubin diagnosis for Johnson Mehl (b) and Voronoi (c) parameterization (from Belhadj et al. (2018))*

Finally, the Bayesian formalism also allows uncertainty to be propagated. For example, when locating earthquakes, generally only the uncertainty on the travel times is considered, although the velocity model (used in the forward problem) has a great influence on the localization result. If this velocity model can be probabilistically inverted with independent data, then it is quite straightforward to propagate the velocity model uncertainty to the event location uncertainty (Gesret et al. 2015) to obtain more reliable localizations whose uncertainties take into account both the picking uncertainty and the velocity model uncertainty. The propagation could certainly be applied to other problems for which the uncertainties of a parameter, usually considered fixed, can be estimated independently.

4.6. References

Agostinetti, N.P. and Malinverno, A. (2010). Receiver function inversion by trans-dimensional Monte Carlo sampling. *Geophysical Journal International*, 181(2), 858–872.

Agostinetti, N.P., Giacomuzzi, G., Malinverno, A. (2015). Local three-dimensional earthquake tomography by trans-dimensional Monte Carlo sampling. *Geophysical Journal International*, 201, 1598–1617.

Andrieu, C., Jasra, A., Doucet, A., Del Moral, P. (2011). On nonlinear Markov chain Monte Carlo. *Bernoulli*, 17(3), 987–1014.

Belhadj, J., Romary, T., Gesret, A., Noble, M., Figliuzzi, B. (2018). New parameterizations for Bayesian seismic tomography. *Inverse Problems*, 34(6), 065007.

Bodin, T. and Sambridge, M. (2009). Seismic tomography with the reversible jump algorithm. *Geophysical Journal International*, 178, 1411–1436.

Bodin, T., Sambridge, M., Rawlinson, N., Arroucau, P. (2012a). Transdimensional tomography with unknown data noise. *Geophysical Journal International*, 189(3), 1536–1556.

Bodin, T., Sambridge, M., Gallagher, K., Rawlinson, N. (2012b). Transdimensional inversion of receiver functions and surface wave dispersion. *Journal of Geophysical Research: Solid Earth*, 117, B02301. doi: 10.1029/2011JB008560.

Bottero, A., Gesret, A., Romary, T., Noble, M., Maisons, C. (2016). Stochastic seismic tomography by interacting Markov chains. *Geophysical Journal International*, 207, 374–392.

Brooks, S.P. and Gelman, A. (1998). General methods for monitoring convergence of iterative simulations. *Journal of Computational and Graphical Statistics*, 7(4), 434–455.

Brooks, S.P. and Roberts, G.O. (1998). Convergence assessment techniques for Markov chain Monte Carlo. *Statistics and Computing*, 8(4), 319–335.

Chiao, L.-Y. and Kuo, B.-Y. (2001). Multiscale seismic tomography. *Geophysical Journal International*, 145(2), 517–527.

Debski, W. (2010). Seismic tomography by Monte Carlo sampling. *Pure and Applied Geophysics*, 167(1/2), 131–152.

Denison, D.G.T., Adams, N.M., Holmes, C.C., Hand, D.J. (2002). Bayesian partition modelling. *Comput. Stat. Data Anal.*, 38(4), 475–485.

Dettmer, J., Dosso, S.E., Holland, C.W. (2010). Trans-dimensional geoacoustic inversion. *The Journal of the Acoustical Society of America*, 128(6), 3393–3405.

Dettmer, J., Molnar, S., Steininger, G., Dosso, S.E., Cassidy, J.F. (2012). Trans-dimensional inversion of microtremor array dispersion data with hierarchical autoregressive error models. *Geophysical Journal International*, 188(2), 719–734.

Dettmer, J., Hawkins, R., Cummins, P.R., Hossen, J., Sambridge, M., Hino, R., Inazu, D. (2016). Tsunami source uncertainty estimation: The 2011 Japan tsunami. *Journal of Geophysical Research: Solid Earth*, 121(6), 4483–4505.

Fichtner, A., Zunino, A., Gebraad, L. (2019). Hamiltonian Monte Carlo solution of tomographic inverse problems. *Geophysical Journal International*, 216(2), 1344–1363.

Gallagher, K., Charvin, K., Nielsen, S., Sambridge, M., Stephenson, J. (2009). Markov chain Monte Carlo (MCMC) sampling methods to determine optimal models, model resolution and model choice for earth science problems. *Marine and Petroleum Geology*, 26(4), 525–535.

Gelman, A. and Rubin, D.B. (1992). Inference from iterative simulation using multiple sequences. *Statistical Science*, 7(4), 457–472.

Gesret, A., Desassis, N., Noble, M., Romary, T., Maisons, C. (2015). Propagation of the velocity model uncertainties to the seismic event location. *Geophysical Journal International*, 200(1), 52–66.

Geyer, C.J. (1991). Markov chain Monte Carlo maximum likelihood. In *Computing Science and Statistics: Proceedings of the 23rd Symposium on the Interface*, Keramidas, E.D. (ed.). Seattle, Washington.

Green, P.J. (1995). Reversible jump Markov chain Monte Carlo computation and Bayesian model determination. *Biometrika*, 82, 711–732.

Green, P.J. (2003). Trans-dimensional Markov chain Monte Carlo. *Highly Struct. Stochastic Syst.*, 27, 179–198.

Haario, H., Saksman, E., Tamminen, J. (2001). An adaptive metropolis algorithm. *Bernoulli*, 7(2), 223–242 [Online]. Available at: http://www.jstor.org/stable/3318737.

Hastings, W.K. (1970). Monte Carlo sampling methods using Markov chains and their applications. *Biometrika*, 57(1), 97–109.

Hawkins, R. and Sambridge, M. (2015). Geophysical imaging using trans-dimensional trees. *Geophysical Journal International*, 203(2), 972–1000.

Hawkins, R., Brodie, R.C., Sambridge, M. (2018). Trans-dimensional Bayesian inversion of airborne electromagnetic data for 2D conductivity profiles. *Exploration Geophysics*, 49(2), 134–147.

Hawkins, R., Bodin, T., Sambridge, M., Choblet, G., Husson, L. (2019). Trans-dimensional surface reconstruction with different classes of parameterization. *Geochemistry, Geophysics, Geosystems*, 20(1), 505–529.

Johnson, W. and Mehl, R. (1939). Reaction kinetics in processes of nucleation and growth. *Trans Aime*, 135, 396–415.

Kijko, A. (1994). Seismological outliers: L1 or adaptive Lp norm application. *Bulletin of the Seismological Society of America*, 84(2), 473–477.

Kirkpatrick, S., Gelatt, Jr. C.D., Vecchi, M.P. (1983). Optimization by simulated annealing. *Science*, 220(4598), 671–680.

Koren, Z., Mosegaard, K., Landa, E., Thore, P., Tarantola, A. (1991). Monte Carlo estimation and resolution analysis of seismic background velocities. *Journal of Geophysical Research: Solid Earth (1978–2012)*, 96(B12), 20289–20299.

Kou, S.C., Zhou, Q., Wong, W.H. (2006). Equi-energy sampler with applications in statistical inference and statistical mechanics. *Ann. Statist.*, 34(4), 1581–1619. doi: 10.1214/009053606000000515.

Loris, I., Nolet, G., Daubechies, I., Dahlen, F. (2007). Tomographic inversion using L1-norm regularization of wavelet coefficients. *Geophysical Journal International*, 170, 359–370.

Malinverno, A. (2002). Parsimonious Bayesian Markov chain Monte Carlo inversion in a nonlinear geophysical problem. *Geophysical Journal International*, 151(3), 675–688.

Malinverno, A. and Briggs, V.A. (2004). Expanded uncertainty quantification in inverse problems: Hierarchical bayes and empirical bayes. *Geophysics*, 69(4), 1005–1016.

Malinverno, A. and Parker, R.L. (2006). Two ways to quantify uncertainty in geophysical inverse problems. *Geophysics*, 71(3), W15–W27.

Mallat, S. and Hwang, W.L. (1992). Singularity detection and processing with wavelets. *IEEE Trans. Inf. Theory*, 38, 617–643.

Menke, W. (2012). *Geophysical Data Analysis: Discrete Inverse Theory*. Academic Press, Cambridge.

Metropolis, N., Rosenbluth, A.W., Rosenbluth, M.N., Teller, A.H., Teller, E. (1953). Equation of state calculations by fast computing machines. *The Journal of Chemical Physics*, 21(6), 1087–1092.

Mosegaard, K. and Tarantola, A. (1995). Monte Carlo sampling of solutions to inverse problems. *J. Geophys. Res.*, 100, 12431–12447.

Noble, M., Gesret, A., Belayouni, N. (2014). Accurate 3-D finite difference computation of traveltimes in strongly heterogeneous media. *Geophysical Journal International*, 199, 1572–1585.

Rawlinson, N., Fichtner, A., Sambridge, M., Young, M.K. (2014). Seismic tomography and the assessment of uncertainty. *Advances in Geophysics*, 55, 1–76.

Ray, A., Kaplan, S., Washbourne, J., Albertin, U. (2018). Low frequency full waveform seismic inversion within a tree based Bayesian framework. *Geophysical Journal International*, 212(1), 522–542.

Robert, C. and Casella, G. (2013). *Monte Carlo Statistical Methods*. Springer, Amsterdam.

Roberts, G.O., Gelman, A., Gilks, W.R. (1997). Weak convergence and optimal scaling of random walk metropolis algorithms. *Ann. Appl. Probab.*, 7(1), 110–120. doi: 10.1214/aoap/1034625254.

Romary, T. (2009). Integrating production data under uncertainty by parallel interacting markov chains on a reduced dimensional space. *Computational Geosciences*, 13(1), 103–122.

Romary, T. (2010). Bayesian inversion by parallel interacting Markov chains. *Inverse Problems in Science and Engineering*, 18(1), 111–130.

Rosenthal, J.S. (1999). Parallel computing and Monte Carlo algorithms. *Far East Journal of Theoretical Statistics*, 4, 207–236.

Rosenthal, S. (2009). Optimal proposal distributions and adaptive MCMC. In *Handbook of Markov Chain Monte Carlo*, Brooks, S., Gelman, A., Jones, G., Meng, X.-L. (eds). Chapman & Hall/CRC, London.

Sambridge, M. (2014). A parallel tempering algorithm for probabilistic sampling and multimodal optimization. *Geophysical Journal International*, 196(1), 357–374. doi: 10.1093/gji/ggt342.

Sambridge, M. and Mosegaard, K. (2002). Monte Carlo methods in geophysical inverse problems. *Reviews of Geophysics*, 40(3), 3–29.

Sen, M.K. and Biswas, R. (2017). Transdimensional seismic inversion using the reversible jump Hamiltonian Monte Carlo algorithm. *Geophysics*, 82(3), R119–R134.

Tarantola, A. (2005). *Inverse Problem Theory and Methods for Model Parameter Estimation*. Society for Industrial and Applied Mathematics, Philadelphia.

Tarantola, A. and Valette, B. (1982). Inverse problems – Quest for information. *J. Geophys.*, 50, 159–170.

Tikhotsky, S. and Achauer, U. (2008). Inversion of controlled-source seismic tomography and gravity data with the self-adaptive wavelet parametrization of velocities and interfaces. *Geophysical Journal International*, 172(2), 619–630.

Wéber, Z. (2000). Seismic traveltime tomography: A simulated annealing approach. *Physics of the Earth and Planetary Interiors*, 119(1), 149–159.

Zhang, F., Dai, R., Liu, H. (2014). Seismic inversion based on L1-norm misfit function and total variation regularization. *Journal of Applied Geophysics*, 109, 111–118.

Zhao, Z. and Sen, M.K. (2020). A gradient based Markov chain Monte Carlo method for full waveform inversion and uncertainty analysis. *Geophysics*, 86(1), 1–79. doi: 10.1190/geo2019-0585.1.

5

Modeling Radar Backscattering from Forests: A First Step to Inversion

Elise COLIN[1] and Laetitia THIRION-LEFEVRE[2]
[1]*ONERA, Université Paris-Saclay, Palaiseau, France*
[2]*SONDRA, Université Paris-Saclay, Palaiseau, France*

5.1. Introduction

A synthetic aperture radar (SAR) image is synthesized from electromagnetic signals captured by a moving antenna. These signals represent the responses of the scene to the radio pulses transmitted by the antenna. The signal within such an image contains valuable information regarding the electromagnetic interactions between the transmitted wave and the illuminated scene. This information relies on various factors, including the characteristics of the observed scene, such as its shape, dimensions, element distribution and materials. Additionally, it depends on the image formation process and the parameters of the observation system, such as frequency, angle of incidence and polarization. The complexity of this information makes it challenging to analyze comprehensively. Its richness becomes even more apparent when multiple acquisition modes are employed, particularly by using various emission and reception polarizations (polarimetry) or two antennas (interferometry).

For a color version of all the figures in this chapter, see www.iste.co.uk/yan/remotesensing.zip.

Inversion and Data Assimilation in Remote Sensing,
coordinated by Yajing YAN.

Extracting information from the observed scene in the image is a challenging task. Typically, this inversion process relies on scene modeling, which involves both the acquisition parameters and the description of the environment. The inversion process, as illustrated in Figure 5.1, aims to recover the descriptive parameters of the environment from the available measurements of physical quantities.

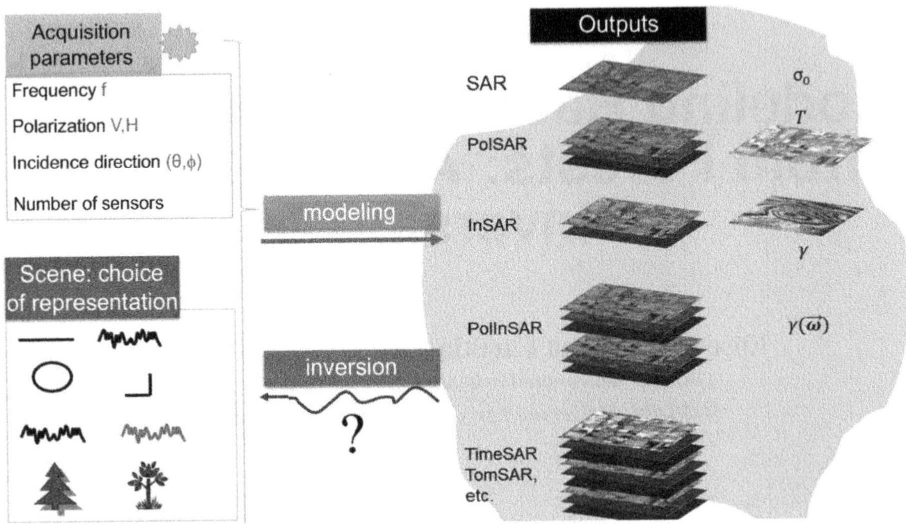

Figure 5.1. *The inversion process: recovering descriptive quantities of the scene from quantities measured by modeling*

The objective of this chapter is twofold: firstly, to provide methodological support for selecting appropriate models based on the parameters to be inverted; secondly, to illustrate that within this framework, the utility of models extends beyond inversion purposes. Modeling both the environment and its responses to a radar signal concurrently leads to a deeper understanding of the phenomenology depicted in the images.

History demonstrates that numerous inversion methods have become feasible thanks to the passage of time and a more distant perspective from the involved phenomenology. The forest environment has been a primary focus of numerous investigations. Consequently, in this chapter, we delve into vegetation. Initially, we offer a historical perspective on how our understanding of radar signals acquired from forests has evolved in tandem with the development of remote sensing sensors.

Next, we will look into the criteria for model selection. For example, one criterion that should be considered is the ability of the model to not only represent the observed scene but also the main interaction mechanisms at play for a given radar configuration. We will also see that the ability of a model to evolve is an interesting characteristic, particularly in a simplification phase. Therefore, by comparing simulated and real radar data, it will eventually be possible to determine which elements are important in this modeling, whether in terms of describing the scene or scattering mechanisms. Moreover, despite that they may contribute thereto, we should bear in mind that not all models have been developed for direct inversion. This point has to be considered when choosing a simulation tool. Modeling can, for example, be used for qualifying algorithms as part of an approach for simulating various acquisition scenarios. It is then possible to calculate the efficiency of these algorithms in a wider range of cases than is possible with the actual measurements available. Finally, modeling can be used to anticipate missions. For example, modeling is useful for predicting the properties of images acquired in bistatic configurations, in new frequency bands, or for new resolution ranges. The choice of model is crucial to meet all these challenges and depends on many contextual factors. We will therefore address these issues, as well as the concomitant issue of model validation.

Finally, for a given and validated model, there are several existing inversion techniques available. We will look into the inversion of one of the most studied key parameters: above-ground plant biomass, that is, the total mass of living vegetation organisms per unit area. We will present the inversion models and techniques presently utilized in the remote sensing community for producing maps that estimate this biomass. To conclude with outlooks, we will propose an extension to this approach to models of urban environments and the description of time series.

5.2. Vegetation model historical background

The history of scattering models is first linked to that of sensors and, of course, to that of computing resources. The primary objective of modeling has always been to gain a better understanding of radar data in order to be able to propose strategies for the inversion of the parameters of the scene under observation. We have chosen the example of radar backscatter from forests. This background will be approached by omitting computer constraints to concentrate on the link between sensor development and model development. It will not be exhaustive but will be able to illustrate the motivations that led researchers to develop these scattering models. In particular, we will not discuss in this background the models that form the core of decomposition theorems based on diffusion mechanisms such as in Lee and Pottier (2017). We will also not discuss the classical computational methods for modeling radar scattering in forests. We refer the reader to the articles cited in this chapter. The reader can also refer to Ulaby and Elachi (1990) or Ruck (1970) and Ruck et al. (1970).

5.2.1. *Evolution of measurements and their understanding*

5.2.1.1. *1980s: the leadership of the Jet Propulsion Laboratory*

During the mid-1980s, NASA initiated a series of missions conducted by its space research center (JPL 1984) for Earth observation[1]. The agency's primary goal at the time was to study the impacts of industrial activities on climate. To this end, imaging radars were installed on board the American shuttle[2], in L-band[3] and in HH polarization (see, for example, the work of Imhoff et al. (1986) or Ranson and Smith (1990)). Simultaneously, field measurements and airborne SAR images were collected with AIRSAR, covering multiple frequency bands (P, L, C) and all polarizations (HH, VV, VH), as detailed in the works of Sader (1987) and Wu and Sader (1987).

These radar measurements, combined with field surveys, have significantly contributed to advancing our understanding of forest radar phenomenology. However, interpreting SAR images remains challenging due to a limited understanding of the interactions between radar waves and their environment. Modeling, therefore, emerges as a valuable tool for comprehending, interpreting and harnessing these images. Initially, modeling primarily utilized the intensity of the measurements, resulting in models where the dominant scatterers were present but without a realistic representation of their spatial distribution, as noted in studies by Richards et al. (1987); Karam and Fung (1988); Durden et al. (1989). These models were developed primarily to enhance our comprehension of the interactions between radar waves and vegetation. By the late 1980s, most new models, such as those developed by Durden et al. (1989) or Ulaby et al. (1990) with Mimics, simulated all of the polarimetric channels, capitalizing on data acquired during AIRSAR airborne campaigns.

5.2.1.2. *The 1990s: the rise of Earth observation satellites*

The 1990s saw the rise of numerous space Earth observation programs that eventually led to the first interferometric measurements[4]: ERS-1 (C Band, VV,

1 Three space missions were conducted: SIR-A in 1981, SIR-B in 1984 and SIR-C in 1994.

2 Hence the name of the mission: SIR for shuttle imaging radar.

3 The central frequency of the band is around 1.2 GHz. The corresponding wavelength in a vacuum is 25 cm, a value well above the size of a leaf or a needle for deciduous and coniferous trees in boreal or temperate environments.

4 The choice of frequency bands was not necessarily the result of a common agreement; for example, the C-band chosen by the ESA for ERS was the result of arbitration between L-band advocates and X-band advocates, as explained in the CCR (2019) report. The Canadian agency followed the ESA's decision for Radarsat-1. As for JAXA, the Japanese space agency chose the L-band to implement forest cover penetration. HH polarization was preferred by the Canadians, as it enables better detection of ships when compared to VV (see, for example, Staples and Rodrigues (2013)).

1991–2000), then ERS-2 (C Band, VV, 1995–2011), Radarsat-1 (C Band, HH, 1995–2013), and JERS-1 (L Band, HH, 1992–1998). These space missions will last several years, and their revisit times will lead to imaging landslides or earthquakes, as with Massonnet et al. (1993) or Zebker et al. (1994). These regular acquisitions paved the way for the first studies of seasonal variations in forests, such as in Proisy et al. (2000), or the study of glaciers with Nakamura et al. (2007) or wetlands by Kwoun and Lu (2009). At the same time, models were being developed to enhance scene representation. This significantly impacted the phase of electric fields, consequently affecting the summation of scatterer contributions. This transition marked a shift from incoherent models, which relied solely on intensity, to coherent models[5], where the resulting phase could be analyzed. These enhancements were necessary to enable the models to accurately provide polarimetric or interferometric phase differences.

Thereafter, numerous other models were created, and in every new model the way the scene is taken into account was further refined, according to the requirements of the applications (e.g. the studies by Chauhan et al. (1991), Sun et al. (1991), Yueh et al. (1992), Saatchi and McDonald (1997) or Lin and Sarabandi (1999)). Improvements to these models led to a better understanding of the phenomenology of interactions between radar waves and forests; therefore, Imhoff (1995) shows with Mimics that the structure of the forest, with constant biomass, had an impact on the level of backscattered intensity. This improved understanding of diffusion has two impacts. Firstly, it enables the emergence of simpler, faster analytical models. In particular, the work by Treuhaft et al. (1996) and Treuhaft and Cloude (1999) led to the appearance of the RVoG (*Random Volume over Ground*) model, which makes it possible to express the interferometric coherence of a forest with just three unknowns (Cloude and Papathanassiou 2003). These direct models can then be used to invert input parameters from radar measurements. The second development marks the emancipation of models from measurements. As such, López-Sánchez et al. (1999) proposed one of the first models of forest bistatic diffusion. With this model, it is possible to anticipate measurement campaigns, with the inherent handicap of not being able to be validated[6].

5.2.1.3. *Since the 2000s: the SAR image explosion*

Since the early 2000s, an increase in the amount of data available has been observed, with a profusion of new sensors in configurations that are either natural

5 Coherent effects refer to phenomena resulting from both amplitude and phase summation.

6 In 2004, DLR and Onera jointly conducted one of the first X-band bistatic SAR campaigns (Dubois-Fernandez et al. 2006). In 2010, Onera and FOI collected bistatic SAR data between 220 and 460 MHz (Baqué et al. 2011).

extensions of older programs[7], either propose new configurations with the development of the X band[8], bistatic data, time series[9] or constellations of micro-satellites[10], which have been able to acquire images with greater agility. Models have become increasingly refined, thus enabling a better understanding of the interactions between electromagnetic waves and forest scatterers. Some studies have focused their efforts on improving the representation of scenes and taking into account mutual interactions, such as Martinez et al. (2000), who utilized AMAPstudio, a software package modeling tree growth developed by Cirad[11] and coupled with the *Branching model* (Yueh et al. 1992).

Still others explored new configurations such as low frequencies (such as Israelsson et al. 2000, Melon et al. 2001 or Nguyen et al. 2006), or bistatic configurations (Thirion-Lefevre et al. 2010b; Everaere et al. 2016). This improvement in models has made it possible to consider the first interferometric simulations with the work of Sarabandi and Lin (2000) and Kobayashi et al. (2000). In addition, the greater availability of radar data is a step further in validating models, as with Thirion-Lefevre et al. (2006), who have worked in radiometry, polarimetry and interferometry. The validation stage is necessary to extend the fields of application of models. The authors have thus used the COSMO model to estimate the attenuation of radar waves through the forest (Thirion-Lefevre and Colin 2007; Lesturgie et al. 2016) or to analyze SAR images in FoPen[12] (Thirion and Colin 2005; Ginolhac et al. 2010; Brigui et al. 2014). Bearing in mind that the physics of interactions has become better understood, models have been simplified, with the emergence of analytical models. The best known to-date is probably the *Random Volume Over Ground model* applied to the inversion of interferometric forest heights (Cloude and Papathanassiou 2003). Since then, many other versions were developed for use in agriculture (as in Lopez-Sanchez et al. 2007, for instance) or for low-frequency observation of forests (Colin et al. 2004). Empirical models were developed, benefiting from the profusion of SAR data (Askne and Santoro 2005, 2009, for instance), whose speed makes inversion possible. These simplification

7 This is the case of Envisat (2002–2012), Palsar (2007–2011) then ALOS 2 (2015-) and RADARSAT-2 since 2007.

8 TerraSAR-X then TanDEM-X for Germany, Cosmo-Skymed for Italy.

9 The launch of the European Union's Copernicus program allows collecting regularly and distributing satellite data free of charge around the globe, and for a wide range of sensors. With Sentinel-1A and 1B, for example, scientists have had access to dual-polarization SAR images since 2015.

10 For example, with Ice Eye since 2018.

11 Available at: http://amapstudio.cirad.fr/.

12 textitFoPen, or *Foliage Penetration*, refers to the whole range of activities that involve penetrating under the forest cover.

choices necessarily implied a restriction of the validity domain. Another solution then consisted of developing metamodels based on the more accurate but slower diffusion models. These metamodels could be seen as transfer functions between the space of input variables and the space of outputs in a diffusion model. A preliminary analysis was necessary to reduce the number of inputs to the variables of interest and to efficiently mesh the input and output spaces (an example of experimental design applied to a forest model can be seen in Thirion-Lefevre et al. 2010a). Finally, this improved understanding of diffusion mechanisms means that only one mechanism can be considered, for example, and then that the inversion of the parameters characterizing it can be performed. For example, the modeling of the double bounce by a dielectric structure of the dihedral type proposed and analyzed by Thirion-Lefevre and Guinvarc'h (2018) has led to the inversion of its permittivity (Couderc et al. 2018; Couderc 2019).

5.2.2. *What should be remembered? And what prospects?*

The history of modeling radar scattering by forests has often closely paralleled the development of sensors enabling the observation of these environments. As this development progressed, new requirements emerged for interpreting radar images. Since the late 1980s, the collective body of modeling work has significantly contributed to enhancing our understanding of the primary scattering mechanisms across different frequencies, the significance of forest structure, the complementary roles of polarization channels and the influence of permittivity on our observations. This improvement has facilitated the development of simpler yet more specialized codes directly applicable to inversion. In addition to these developments, new tools have emerged, including experimental designs, which remain underutilized in the field of remote sensing, and learning algorithms, which can now be employed for more extensive model inversion.

These latter then serve to provide controlled terrain data (model inputs) associated with observed radar quantities (model outputs). In addition, this knowledge of scene description and the flexibility of the radar configurations that can be simulated make scattering models invaluable tools for testing processing algorithms, some of which are dedicated to inversion. For example, the European Space Agency has included a forest SAR image simulator developed by Williams in the free tool PolSARpro-Sim[13].

It is probably the development of sensors that will continue to influence the development of models. With time series, for example, we are prompted to redefine the quantities observed and study the impact of time on their variations (see, for

13 Available at: https://dragon3.esa.int/documents/653194/662629/PolSARproSim_Design.pdf.

example, Weissgerber et al. 2016; Paillou et al. 2020). Besides phenomenological and statistical considerations, the huge amount of images now available also raises the question of how should they be processed. Learning algorithms can be seen as ideal black boxes; the characteristics of the situation being observed can actually be inverted without the need for modeling. Is this really the case? These questions are likely to shape the future direction of modeling research.

5.3. How to choose a model for inversion?

The background history of forest diffusion models has shown us how model development first responded to the need to understand images in acquisition modes that have become technologically accessible. Although inversion has not always been the main purpose of models, an increasingly detailed understanding of the diffusion mechanisms that occur has nevertheless been a preliminary basis for the first inversions, via for example, simplified models. As a result, we are now going to see that the choice of a model already depends on how its use is conceived.

5.3.1. *Different inversion approaches*

There are two main categories of inversion processes, as schematically illustrated in Figure 5.2. In a black-box approach, where the internal physical operation is not described, the procedure consists of inverting the transfer function of the model without any knowledge of its physical structure. In a model where the physical interactions are described by the model, this direct inversion of the model transfer function is still possible. However, this inversion scheme still remains uncommon, even when it is known that there exists solutions such as metamodels, model simplification or learning algorithms trained with these models. Generally, the proposed inversion schemes result from the phenomenological analysis of model outputs; analysis and inversion are then jointly developed.

We now look into the choices underlying this second category of models, those whose inner workings can be controlled.

First of all, the target object has to be defined for the modeling process, namely the object to which the measurement will give access. This may be a radar image, with or without taking into account the formation of the image by a SAR process. It may also concern the electric field resulting from the interaction of the emitted wave and the modeled scene, possibly for a given geometric configuration and/or a particular wave frequency. This target object can also be any derived radar quantity, such as interferometric coherence or the backscatter coefficient.

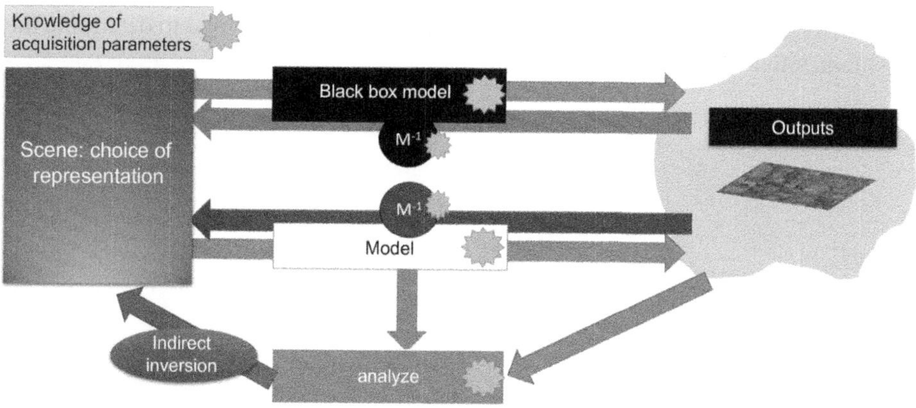

Figure 5.2. *The different inversion approaches to modeling: "black box"-type inversion or model inversion making phenomenological analysis possible*

Relatively to the model, the choices then concern both the representation of the medium under observation, as illustrated in Figure 5.3, and the existing computational methods or wave–object interaction mechanisms deemed useful and that should be taken into account to provide the targeted simulation output. For example, as elementary building blocks, an ellipsoid (Figure 5.3(a)) or a revolution cylinder (Figure 5.3(b)) could be used. Maxwell's equations can also be solved for a mesh of the object to be modeled in 3D (Figure 5.3(c)); diffusion can also be considered for a cloud of particles, as in radiative transfer codes (Figure 5.3(d)). A tree can be modeled by a full mesh, facilitating the formulation of an integral field (Figure 5.3(e)), by cylinders distributed in layers and in a fixed canopy shape (Figure 5.3(f)), by connected cylinders and ellipsoids, or even an ultra-realistic 3D model of plant architecture (Figure 5.3(h)) (Jaeger and De Reffye 1992). Different types of mesh will make the computation of attenuation possible (Figure 5.3(j)), different trunk distributions will enable the study of the effect of the ordered or random distribution of trees (Figure 5.3(k)), and the inclusion of slopes will enable visualizing the influence of terrain effects (Figure 5.3(l)) (Villard et al. (2010)).

We will now see that these two types of choices depend on the expected purpose. Models can indeed be dedicated to proposing a direct inversion function, but also to types of functions other than pure inversion, such as anticipating acquisition campaigns, producing annotated data for deep learning, developing specific strategies for certain functionalities such as target or anomaly detection, or even for qualifying algorithms. It should be noted that, in all case studies, a model designed for data inversion must meet computation time constraints, which are highly dependent on the number of model input and output parameters. In order to optimize model complexity, experimental designs or learning algorithms can be used to determine the most influential model inputs.

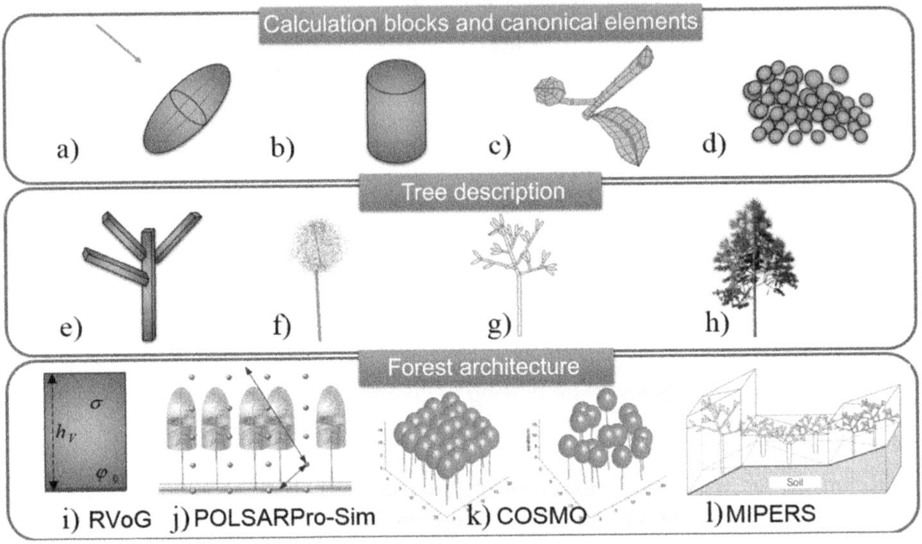

Figure 5.3. *Different choices to be made for modeling: (ad) examples of elementary computation bricks and canonical elements; (e–h) examples of tree descriptions; (i–l) examples of forest architectures*

5.3.2. *Choosing according to the intended purpose*

5.3.2.1. *Qualification of algorithms*

When the performance of an algorithm should be anticipated, the model feeds into a simulation, which is understood in a broader sense and includes the description of a scenario, sensor parameters and a particular acquisition configuration. Once the modeling principle has been established, it is possible to run several simulations following a variety of acquisition configurations, thus providing a wealth of test data for verifying the performance of the algorithms developed and establishing a parametric study of this performance (for example, in Lebarbu et al. 2021). In this type of study, high computational performance is expected, and to propose large scenes, resorting to statistical simulations is often considered to capture large homogeneous natural areas. In the context of algorithm qualification, the PolSARpro tool should also be mentioned. Developed under the umbrella of ESA, it provides, in addition to real data, sets of SAR images of forests simulated with PolSARpro-SIM (Williams et al. 2007).

5.3.2.2. *Data production for deep learning*

In the same vein, modeling plays a crucial role in making available data along with ground truth. Deep learning algorithms, which are currently in full expansion, require, when supervised, the availability of labels or objectives to be achieved in association with training data. In remote sensing, more than the abundance of image-type data, it is the disposition of these labels that is most often missing. For example, to train a network for plant species classification, it is necessary to have remote sensing images as well as an overlay layer describing the plant classes contained in the image. If a modeling tool that makes this possible is available, then the possibilities for supervised learning are multiplied. Since modeling can never rigorously replace real measurement, combining learning from real data and learning from modeled data, as in Yokoya et al. (2020), could become an attractive alternative, although it remains uncommon in the literature for the moment. Here again, computational performance is a key criterion.

5.3.2.3. *Anticipation of future campaigns*

As a consequence of the constant evolution of sensor technology, scientists are constantly faced with acquisitions achieved with new configurations. Images are acquired in new frequency bands, reaching previously unattainable resolutions. These are sometimes calculated using innovative geometric acquisition modes, such as circular trajectories or even bistatic configurations. Experience shows that each new acquisition mode often brings unexpected results when interpreting the information obtained from the images thus produced. Therefore, TerraSAR-X resolutions of about a meter have contributed to understanding the extent to which it was impossible to consider urban environments as deterministic environments in X-band, as long as estimates are used in a spatial neighborhood (Weissgerber et al. 2016), since two adjacent pixels contain objects that are different at the wavelength scale. Acquisitions made along circular paths in the X-band, intended to achieve unrivaled spatial resolutions, have demonstrated both the extreme anisotropy of electromagnetic behavior in this band and the inability to guarantee the isotropy assumption to achieve the expected resolutions (Cantalloube and Colin 2008).

It is therefore tempting to use modeling to simulate new configurations, anticipate the associated difficulties and verify the expected benefits.

It was this approach that motivated the study of bistatic scattering by forests, for which experimental data are still lacking. Therefore, the aim of the thesis work in Everaere (2015) was to anticipate real P-band bistatic measurements on the forest, either for inversion of physical forest parameters or detection of artificial objects under the canopy. However, the problem of validating the COBISMO bistatic model emerged, and the proposed solution was to use polarimetric measurements achieved at the optical scale on carbon nanotube forests (Everaere et al. 2016). For the time being, it is not possible to compare these measurements with real data.

While it may be unrealistic to fully anticipate all new situations, models can provide valuable information that allows experimenters to better plan their measurement campaigns. For example, Thirion-Lefevre and Guinvarc'h (2018) and Thirion-Lefevre et al. (2020) have demonstrated that scenes containing dihedral structures can potentially exhibit very low VV polarization responses. This can result in anomalies in the polarimetric responses of such scenes, as discussed by Brisco et al. (2015). Therefore, this information must be taken into account when planning acquisition campaigns and inversion procedures.

Mission anticipation is undoubtedly the application area where the selection of models remains the most open.

5.3.2.4. *Sensing*

Forest models can be inverted to detect targets concealed beneath plant cover, as proposed by Cloude et al. (2004). When the recorded signal is a composite of the desired target and the forest environment, understanding how to model the forest's response allows for the separation of the two signals. In Cloude et al. (2004), the response of a random volume is subtracted, and in Colin et al. (2005), the response of a "trunk with random volume" model is utilized to improve the contrast between a target and its surroundings.

Similarly, modeling canonical responses can play a role in the development of SAR algorithms designed for target imaging, as shown in Brigui et al. (2013). In this case, the elementary scatterer model departs from that of an isotropic white point and instead adopts the form of a diversely oriented metal plate. This results in the creation of a SAR image that enhances target detection.

5.3.2.5. *Semantic segmentation*

Most classification approaches for polarimetric SAR images (PolSAR) are based on different polarimetric decompositions of the target. These classifications are very similar to semantic segmentation, since the goal is to semantically understand the role of each pixel in the image. These polarimetric decomposition methods can be seen as a form of modeling, as they allow the extraction of many descriptive parameters. For example, in the method proposed by Cloude and Pottier (1996), which is one of the most widely used approaches, polarimetric information is converted into three parameters (entropy, double bounce angle and anisotropy), each of which is then associated with a particular physical interpretation. The characteristic space formed by these three parameters can be divided into regions corresponding to distinct scattering behaviors.

This work has been supplemented and improved by numerous subsequent contributions. The decomposition of Freeman and Durden (1993) models the covariance matrix as the contribution of three diffusion mechanisms: surface diffusion, double-bounce diffusion and volume diffusion. Yamaguchi et al. (2005) introduced a fourth helix component into the model to reflect non-symmetric effects. Touzi (2016) provided a state-of-the-art review of all these existing decompositions. The trend in the current literature on polarimetric decompositions is toward increasing the complexity of the decompositions proposed, hoping to best make use of the potential wealth of information contained in the polarimetric matrix (Maghsoudi et al. 2013).

5.3.2.6. *Better understanding of phenomenology*

Polarimetric decompositions play an important role in understanding the existing phenomenology, since they are often intended to give meaning to pixel measurement.

However, many other models greatly contribute to understanding phenomenology. Model modularity is an asset for properly understanding the contribution of scatterers or mechanisms in the full response of a scene. With this modularity, the first possible approach consists of starting with the simplest possible model and of progressively increasing its complexity. To model a forest, for example, the behavior of a dielectric cylinder that acts as a descriptive building block for a branch or trunk will need first to be understood. Once the elementary behavior of this object is well understood, an assembling stage involving various objects can begin to describe an increasingly realistic forest, as in the contributions by Karam and Fung (1988); Karam et al. (1992); Sarabandi and Lin (2000). Conversely, another possible approach consists of starting with a model that describes a tree or forest as completely as possible and then removing scatterers or mechanisms that prove to be negligible in the radar configurations under consideration. This was the approach adopted in Thirion-Lefevre et al. (2006), where the contribution of leaves was finally removed for application to P and L bands.

Modularity is often associated with approximate models, which generally do not take into account the mutual interactions between scatterers. A trade-off between modularity and computational accuracy has thus to be achieved. However, a few examples of a hybrid approach can be found, as in Bellez et al. (2011). In addition, approximate models are often scalable with a view to answering other questions, such as the influence of ground slopes, tree distribution, or density on attenuation effects. Finally, within the framework of this analysis, it is essential to know whether the model can be trusted, and therefore to define its validation domain.

5.3.3. *Validity and validation domain*

One of the crucial questions when choosing a model concerns the level of trust it can be granted. A given model makes decisions regarding the description of the environment and how physical interactions should be computed. A model domain of validity refers to the set of assumptions under which it can be applied and has been validated[14].

One of the most frequently encountered assumptions is the order of magnitude of the frequencies utilized by the sensor or, equivalently, that of the wavelength. In fact, environment modeling must be considered at spatial scales of the same order of magnitude as these wavelengths. Typically, for a wavelength of 1 m, a tree can be roughly described by its trunk and main branches; for finer wavelengths, the foliage must be taken into account. A trunk can be modeled by a cylinder, whose canonical response can be simplified if its height is greater than the wavelength. Other potential restrictive assumptions include those relating to electromagnetic properties. In fact, certain simplifications can be considered for some ranges of dielectric permittivity, and some models will only apply to metallic objects.

First of all, we should note that it is important to be careful with the simultaneous use of several boundary cases. For illustration purposes, an example highlighted in Trouve et al. (2011) concerns the computation of bistatic scattering by a dielectric sphere, whose radius is small compared with the wavelength. The expression of the polarimetric matrix obtained is not the same as for a perfectly conducting sphere. This may seem counterintuitive, as the metallic case is often considered to be the limiting case of a dielectric sphere for which the imaginary part of the permittivity has purposely been made to tend to infinity. Nevertheless, the case of Rayleigh scattering is already a limiting case where the radius has been made to tend toward zero.

Therefore, the combination of several restrictive conditions makes the determination of validity ranges particularly complex. This difficulty also comes across when considering multi-scale environments such as forests. For this reason, it is necessary that a validation strategy be implemented. Generally speaking, this strategy consists of comparing simulation results with measurements performed in

14 Bearing this observation in mind, it should be noted that, in many books addressing electromagnetic modeling, we will most of the time have access to calculations or formulations that are restricted to the authors' application. For example, analytical developments often assume a monostatic, far-field configuration. When results are illustrated, they are often in terms of intensity, at best for co-polarizations. On the other hand, the phase is often hardly plotted. It is therefore important to keep in mind that the model published has been so within a given context, for a given application, and validated under these conditions, and that we should verify that this model corresponds to our application and that it is valid for this specific application.

the same acquisition configuration. In the absence of data, this involves at least stating the expected physical principles and verifying that they are respected by the simulation.

Mastery of the scene often requires a detailed description of this scene, according to the criteria identified as being important for modeling. We reconsider the example of a P-band forest model intended to serve as the basis for biomass inversion. If the model is designed to model trunks as cylinders with an inclination, diameter and height, it is necessary to have a description of the scene measured using these same criteria. This description of the environment is known as ground truth. Obtaining or fine-tuning this ground truth constitutes a scientific topic on its own, and still relevant: recently, Nagai et al. (2020) stressed the importance of collecting field data at several sites in order to improve our still inadequate understanding of satellite data.

One way of getting around the difficulty of obtaining ground truth associated with real measurements is to carry out measurements in a reduced-scale anechoic chamber. The underlying idea is to reconstitute a controlled scene. The description then becomes easier (Bellez et al. 2011).

A second strategy that avoids the need for ground-truth measurements consists of resorting to cross-validation of different models. This approach typically involves comparing the results obtained by an approximate model, based on several simplifying hypotheses, with those obtained by rigorous methods, based on the exact resolution of the electromagnetic system under consideration.

Finally, the absence of suitable real data sometimes prevents validation. This is particularly the case when trying to model images in future technological configurations, for instance, when trying to predict the content of future images in resolution ranges that do not yet exist, or in particular geometric configurations. The most illustrative example is that of the modeling of polarimetric bistatic data.

Once the validation strategy has been determined, it will then be necessary to choose the type of data to be compared. Depending on the model outputs upon which the subsequent inversion will be based, not only will the radiometric outputs, according to the use cases, have to be verified, but also potentially the polarimetric, interferometric and statistical quantities as well as other ones. In particular, if the model is a coherent one, the phase of the signal obtained will be an essential element of validation, enabling the model to be used in advanced modes.

5.3.4. *Summarizing the choice of model*

As we have seen, the choice of a model is based on both the way in which the scene is described and the physical way in which electromagnetic calculations are perceived. The choice depends on the inversion strategy and the intended purpose. The first step is therefore to determine whether the phenomenological analysis provides added value to the target application and whether other characteristics are relevant, such as computational time, modularity and accuracy. Finally, the calculations and description chosen will have to be adapted to the target configuration to ensure that the model is valid for the application scenario.

5.4. Biomass inversion

The preceding background and its proposed reflections on how to choose a model and ensure its validity have provided us with a better insight into the modeling requirements for radar observation of forests. We now propose to focus on biomass inversion and identify the contributions that modeling has made in this field.

5.4.1. *Challenges*

In ecology, biomass is the total mass of living organisms in a given environment at a given time. It is estimated in terms of mass per unit area or volume, rather than absolute mass. Very often, it is implicitly meant that we are referring to plant biomass. A distinction can also be made between dry biomass and wet biomass: the former corresponds to the organic mass after evaporation of the water it contains. It is generally preferred to the notion of wet biomass since it is a notion that abstracts from water concentrations, which may fluctuate depending on the time of measurement. Finally, it should be pointed out that we are only interested in above-ground biomass, as opposed to that which also includes root volume. Dry above-ground plant biomass is a key parameter that appears for the study of climate since forests constitute the main sources or sinks of carbon. In 2013, the European Space Agency selected the Biomass mission to become the 7th Earth exploration mission[15]. In particular, this mission should allow for better understanding of the contribution of forests to the carbon cycle through a global estimate of forest biomass, but also through the study of forest disturbances such as deforestation. In the latter case, the aim is to assess changes in biomass. For more details on this forthcoming mission, we refer the reader to Quegan et al. (2019) for a comprehensive article on the subject.

The study of biomass has mobilized many researchers around the world, and its inversion is a very complex subject, as it covers several areas of SAR observation. The first of these is radiometry, with the correlation between biomass and backscatter

15 Available at: https://www.esa.int/Newsroom/Press_Releases/ESA_s_next_Earth_Explorer_satellite.

coefficients having been widely exploited for biomass inversion. The second is interferometry (and its derivatives, namely polarimetric interferometry, and tomography), with mean forest height, which can be related to biomass. We will complement these first two techniques with two other, more marginal, approaches that can however also contribute to biomass inversion.

5.4.2. *Regression of backscatter coefficient curves*

Classically and historically, the first biomass inversions were based on SAR radiometric images of forests. The technique consists of plotting the backscatter coefficient σ_0 curve according to biomass, based on images for which joint biomass measurements are known. A parametric regression of this curve results in inverting biomass from new values of σ_0. Studies carried out in P-band have shown that HV polarization gives the best correlation results. Quegan et al. (2019) recall the relation between biomass W and polarization backscatter HV:

$$\gamma^0_{HV} = cW^p \qquad [5.1]$$

with $\gamma^0_{HV} = \sigma^0_{HV} \cos^{-1} \theta_l$, where θ_l is the local angle of incidence, and c and p the parameters to be determined by regression. The limitations of this inversion lie mainly in the fact that these backscatter/biomass curves depend greatly on the type of forest being observed. Moreover, there is a phenomenon of backscatter saturation for high biomass, leading to significant estimation errors in this area (see the synthesis achieved by Papathanassiou et al. 2021). The values of parameters p and c can therefore vary according to forest type and sensor parameters. Consequently, the inversion model is not the same for a tropical forest as for a boreal forest, mainly due to the differences in backscatter levels between these two vegetation types. In addition, this inversion is highly sensitive to topography and humidity, which lead to a dispersion of backscatter values, as specified by Quegan et al. (2019).

Quegan et al. (2019) also points out that the physical explanation for this relation is not well understood. In fact, the greater correlation of this HV polarization with biomass is to be compared with the literature. For the lowest frequency bands, it was shown that forest phenomenology is mainly linked to trunk phenomenology (see, for example, Thirion-Lefevre et al. 2006). However, on flat ground, the response of trunks, whose orientation is almost vertical, is dominated by signals acquired in the HH and VV copolar polarizations. It is also known that multi-scattering is favored by HV polarization and, in forests, this is mainly associated with branches. We therefore need to understand the configurations in which these multi-scattering are privileged, and how this might explain why biomass is more sensitive to HV at low frequencies.

Finally, these regression techniques are based on in situ biomass measurements, yet the available data are still too scarce to consider a worldwide inversion (Quegan et al. 2019).

Initially, radar measurements combined with field measurements were used to evaluate the robustness of this relationship between the backscatter coefficient and biomass. This is therefore an empirical model. To go further with this inversion, it will probably be necessary to resort to modeling in order to better understand the nature of the interactions between radar waves and the building blocks of forests at low frequencies. These interactions are likely to differ according to the type of forest under consideration: boreal, temperate or tropical. Corrections for topography and humidity effects can also be tested on forest scattering models, with the caveat that these models are probably better controlled for boreal and temperate forests than for tropical forests, whose description is more difficult. It is also conceivable to use these models for producing synthetic data on boreal or temperate forests to serve as input for regression algorithms. Finally, it appears that a major area for progress would be the modeling of diffusion data from tropical forests. The contribution of the Biomass mission, combined with the time series of Sentinel-1, will certainly make it possible to improve the knowledge of these environments, which are still poorly documented.

5.4.3. *RVoG model in PolInSAR*

Another possibility for inverting biomass is based on the knowledge of the average height of forests.

The advent of polarimetric, interferometric and now tomographic sensors has led to considering the inversion of this height based on knowledge of the phase center of the forest under observation. The difficulty lies in the fact that the position of this phase center can vary from the ground to the tree tops, depending on radar parameters (frequency, polarization) and forest characteristics (such as cover density, branch orientations and moisture content of scatterers). In interferometry, most height inversion techniques are based on the use of a descriptive forest model in the form of a volume of randomly oriented particles: the RVoG model[16] (Treuhaft et al. 1996). As shown in Figure 5.4, this model takes three parameters as input: the soil interferometric phase φ_0, the total volume thickness or vegetation height h_V and the

16 As can be inferred from its name, the RVoG model (*Random Volume over Ground*) assumes a random volume. This implies in particular that the radar wave does not penetrate beyond the tree crown, which is generally the case for short wavelengths such as in the C band and beyond. This also naturally depends on the density of forest cover. The idea remains that the random volume assumption is acceptable as long as the contribution of the trunks is not visible. In practice, this can be verified if almost an identical backscatter intensity in VV and HH can be observed.

volume attenuation σ. According to the RVoG model, the generalized PolInSAR coherence shape represented in the complex plane should have the shape of a line segment, connecting the interferometric center corresponding to the bare ground $e^{i\varphi_0}$ and the volume interferometric coherence $e^{i\varphi_0}\gamma_{veg}$. From the linear regression of this shape, some or all of the three descriptor parameters can be inverted using various techniques, as summarized in Papathanassiou et al. (2021).

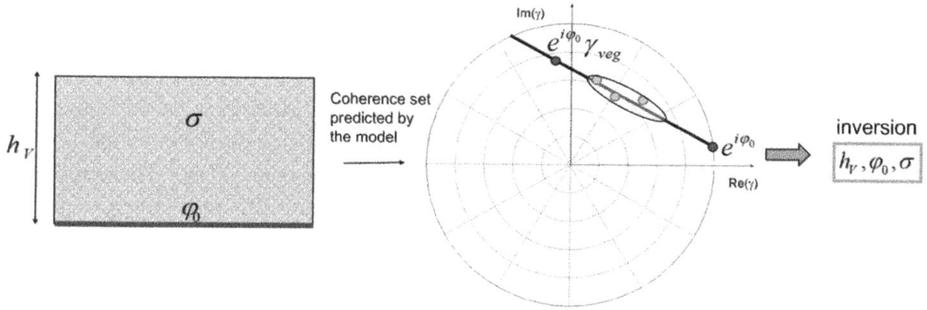

Figure 5.4. *The RVoG model inversion: the description of vegetation as a random volume is represented by a generalized interferometric coherence modeling based on three parameters*

Once the height is obtained, through the use of allometric equations, the biomass can be estimated. The main limitations of this model are linked to the simplified description of forest volume, which is poorly suited to frequencies that are too low or, on the contrary, too high. Several approaches have been proposed to overcome these limitations. At low frequencies, Colin et al. (2005) proposed the addition of a trunk layer, and Lopez-Sanchez et al. (2007) proposed a preferentially vertical orientation component in the modeling. Colin et al. (2005) also proposed including an a priori on attenuation in the inversion procedure. It should be noted that inversion schemes have also been proposed for partial polarimetry-based configurations (Romero-Puig et al. 2021).

Should the soil contribution be eliminated to isolate volume scattering and improve the inversion of the phase center's height? How can decorrelations due to non-simultaneous acquisitions be compensated for? When considering these topics, should we adopt the same strategy for C-band (e.g. Sentinel-1) as for P-band (e.g. Biomass)? Once again, forest structure is likely to play a major role, as demonstrated by Imhoff (1995) using the Mimics model.

5.4.4. *Approximate model inversion*

So far, we have addressed the two main approaches to biomass inversion that have been predominantly studied. The first is based on an empirical model that is not yet fully understood and needs to be improved under certain conditions. The second is based on an analytical model that makes it possible to invert not the biomass directly but the average forest height, which can then be used to calculate biomass. This model is based on approximations that partly limit its scope of validity, but on the other hand, it is simple and fast enough for integration into an inversion loop.

Nonetheless, it is possible to take into account other strategies, such as the use of approximate descriptive models such as Cosmo (Thirion-Lefevre et al. 2006), PolSARpro-Sim (Williams et al. 2007) or Mipers (Villard et al. 2010). These models are based on describing a forest as a set of cylinders and ellipsoids. Each of these elements is associated with canonical electromagnetic mechanisms: simple diffusion, double rebound and triple rebound for cylinder–soil interactions. The calculation of forest scattering is then seen as the sum of the attenuated contributions of the scatterers in the scene. These two tools therefore take a large number of parameters as input, far in excess of the number of observables. As such, they are not suitable for inversion. It is however possible to propose alternative inversions.

For example, an inversion technique for tree heights from purely interferometric data was developed using the Cosmo model (Thirion-Lefevre and Colin 2007). This technique was built upon an empirical relationship, validated on the basis of simple models and real data, between interferometric phase center height, attenuation and total volume height. It was shown that when the main scattering takes place in the canopy[17], it is then verified that:

$$h_{foret} = h_{interf} + \frac{\cos\theta}{2\alpha} \qquad [5.2]$$

where h_{interf} is the phase center height at m, θ is the angle of incidence and α is the linear attenuation coefficient in Np/m. It should be noted that for canopy scattering, volume scattering applies, and, as such, attenuation and phase center heights are similar for HH and VV in this case. If this is not the case, this is due to double trunk-to-ground bounces that play a significant role.

This is a typical example of an indirect inversion scheme, corresponding to the last case illustrated in blue in Figure 5.2: it is the analysis enabled by the model that has

17 This is observed empirically by almost identical values between the backscattered intensities in VV and HH.

driven us to propose an empirical relationship between these parameters, as well as an associated inversion scheme.

5.4.5. *Metamodels*

We have seen that inversion methods are often the result of a trade-off between a simple, often empirical model, which allows inversion to be considered without any a priori knowledge of the physical interactions, but whose scope of validity is often restricted, and a precise model however often impossible to invert as such.

Another approach consists of developing a metamodel; in other words, a black box, which would link inputs (e.g. forest and radar configuration descriptors) to outputs (e.g. backscatter coefficients and/or attenuations and/or interferometric coherences for all polarimetric channels). The main issue with these metamodels lies in their efficient construction, in order for the output space to be regularly described with a minimum of different configurations for the inputs. This requires the implementation of a strategy that allows correctly choosing the mesh of input space data. This is known as kriging or experimental design (see, for example, Bilicz 2011). Input data may originate from measurements but also from diffusion models. Once built, the metamodel is as fast as a simple model and can therefore be used for inversion. Its scope of validity, and therefore of application, will remain that of the initial model used to generate the input and output data. If it originates from real data, it will remain restricted to the domain defined by these measurements.

An example of the effectiveness of this metamodel approach is illustrated by Thirion-Lefevre et al. (2010a), who proposed an experimental design to develop a Cosmo metamodel. In their exploratory study, allometric relations were used which resulted in reducing the number of inputs by inferring certain descriptive parameters of the forest from its age. The study was restricted to low frequencies; as such, only trunks were considered for simulating trees. Eight outputs, backscatter coefficients and interferometric height for all polarization channels, as well as attenuation for copolar channels, were generated for a pine forest measuring 200 m by 40 m on the ground, for incidence angles from 20° to 70° in 1° steps, and for frequencies ranging from 350 MHz to 2 GHz in 10 MHz steps. In all, more than 8,000 simulations and several days of computing time would have been needed with their Cosmo model. The metamodel provided all these outputs in just a few minutes.

The metamodel was then applied to a test case in Vasko et al. (2011) for the joint inversion of soil moisture and forest age (and thereby its mean dimensions) from backscatter coefficients in VV, HH and HV. In addition, it is possible to introduce uncertainties in the measured data and represent them, by way of the metamodel, by confidence intervals on the inverted outputs.

The use of metamodels seems promising for the inversion of quantities such as biomass, especially as it is then easy to take into account the uncertainty in radar measurements. The development of a metamodel however comes at a cost: the input space must be correctly sampled, which requires a large amount of high-quality field data.

5.4.6. *What should be remembered and expected?*

To conclude this section on biomass inversion, we can observe that this quantity is not systematically inverted directly from the measurement, all the more that there are no observational parameters that make it possible to unequivocally estimate it: forest structural parameters almost always impact radar observables.

It is therefore often necessary to address other types of parameters (vegetation height, density and structure) and then link biomass thereto by way of allometric equations.

The trend is therefore toward the complexification of forest representation models to better understand the impact of its structural parameters. We can also consider improving inversion strategies, as illustrated by the utilization of metamodels, which make it possible to take uncertainty into account in measurements.

The most recent work dedicated to deep learning seems promising, given the performances observed in other fields related to *big data* (Zhu et al. 2017). The significant increase in the number of SAR measurements available suggests that these techniques could also be used in radar remote sensing. For supervised learning methods, it is the lack of sufficient ground truth data that hinders the use of these techniques for inversion; it is nonetheless possible to bypass this difficulty through the use of other types of data as ground truth, as carried out by Brigot et al. (2019) with LIDAR data. For unsupervised learning methods, the promising results of Di Martino et al. (2021) brought some momentum to the use of these techniques for inversion.

5.5. Conclusions and outlook

In this chapter, we have proposed the main guidelines for choosing an inversion model and method using forests as an example.

The historical background has allowed us to highlight how the development of models has undergone closely interwoven with technological developments in sensors and a better understanding of the phenomenology of available measurements. The choice of the model, therefore, depends on the need for this better phenomenological understanding, as well as on the intended purpose of the inversion. We have illustrated these principles with biomass inversion, the most widespread use case of models in SAR imagery.

There is a large number of other areas where these principles can be applied, such as the ocean, agricultural areas, the cryosphere and urban areas.

The most incipient area of modeling, and consequently of inversion, is probably that of urban areas, as most applications are done at the scale of buildings; the resolution ranges required are therefore more demanding, and only relatively recently have they been reached at the scale of civil satellites, in particular with the launch of the TerraSAR-X sensor in 2007. On the other hand, these improvements in resolution are often accompanied by the use of higher frequency bands. The interactions involved become all the more difficult to understand and model. Moreover, these are generally inhomogeneous environments in terms of scatterers, regardless of their size, orientation, or materials therein.

Applications of interest in urban areas particularly include the classification or building extraction, and the inversion of their height, materials and facade orientation. The production of classification maps is a response to a need for land-use planning and management. In this context, the "building" class is of particular interest.

Urban areas are known to appear as particularly strong zones in purely radiometric images, but this intensity criterion is not enough to achieve interesting classification performance. In polarimetry, the first approach consists of modeling the building as a dihedral. However, the corresponding canonical target, the "metallic dihedron oriented parallel to the trajectory", is a mere flight of fancy. Most of the time, there is no indication that the walls are parallel to the trajectory or that the materials that they are made of are conductive. Finally, this approach ignores the countless details that comprise a facade, and which become important as frequency increases.

Today, a series of works are emerging that utilize modeling to enhance our comprehension of these environments. This involves extending the canonical dihedral model by considering disorientation effects, as explored in Li et al. (2016) and Atwood and Thirion-Lefevre (2017), as well as examining the dielectric properties of the materials constituting the dihedral, as discussed by Thirion-Lefevre et al. (2020).

This approach has recently provided insights into the double Brewster effect, which illustrates how the materials constituting a dihedral can induce a significant disparity between the intensities received in the VV and HH channels. This effect, while it may introduce interpretation challenges in polarimetric SAR images, also offers the opportunity to discern specific material characteristics.

As such, Couderc (2019) proposed the inversion of dielectric permittivities from the measurement of dielectric structure backscattered fields. This approach could have applications such as the detection of structural defects or the estimation of water content in soils or concrete.

In addition to determining the overall orientation of a building, it can be valuable to ascertain the orientation of the primary scatterers within the resolution cell. In Weissgerber et al. (2017), the authors explore the inversion of facade or component orientations relative to the acquisition trajectory using radar polarimetry. This estimation is greatly facilitated by employing time series data, allowing for the precise extraction of scatterers with deterministic responses.

Time analysis is playing an increasingly crucial role in the contemporary remote sensing landscape. The advent of the big data era in remote sensing has significantly facilitated access to images from around the world. Beyond the multimodal nature of these data, the temporal dimension represents a substantial shift in processing configuration. This is particularly evident in SAR imagery, where the inherently random nature of speckle signals necessitates considering multiple statistical realizations of the same measurement to derive accurate estimates. Until recently, due to data limitations, these estimates were primarily produced in a spatial context. Nowadays, they can be readily generated by capitalizing on the temporal redundancy while preserving spatial resolution. In the specific case of urban areas, this aspect is fundamental for a comprehensive understanding of radar phenomenology.

However, in both cities and forests, it is important to understand the temporal behavior of these measurements, either when no major change happens in the scene, or when modifications occur over time. To-date, our knowledge of radar phenomenology is mainly built from the results of the analysis of sporadic acquisitions. With these time series, many questions have been raised that could change our understanding of SAR images. What amplitude statistics do the time profiles of a SAR image follow over a speckle area or a bright spot? How does the temporal coherence parameter of acquisitions behave? Are two different forest realizations totally decorrelated between two acquisitions? What biophysical parameters are the signals sensitive to? Do the signals depend on meteorological conditions or on the geometric acquisition conditions in relation to the structures in question? Work in this field is still in its early stages, and without a doubt, it can be asserted that modeling will be an essential tool for finding answers and advancing inversion further.

5.6. References

Askne, J.I.H. and Santoro, M. (2005). Multitemporal repeat pass SAR interferometry of boreal forests. *IEEE Transactions on Geoscience and Remote Sensing*, 43(6), 1219–1228.

Askne, J.I.H. and Santoro, M. (2009). Automatic model-based estimation of boreal forest stem volume from repeat pass C-band InSAR coherence. *IEEE Transactions on Geoscience and Remote Sensing*, 47(2), 513–516.

Atwood, D.K. and Thirion-Lefevre, L. (2017). Polarimetric phase and implications for urban classification. *IEEE Transactions on Geoscience and Remote Sensing*, 56(3), 1278–1289.

Baqué, R., Dreuillet, P., du Plessis, O.R., Martineau, P., Ulander, L.M., Gustavsson, A., Jonsson, T., Stenström, G. (2011). Results of the LORAMbis bistatic VHF/UHF SAR experiment for FOPEN. In *Proceedings of 2011 IEEE CIE International Conference on Radar*, Chengdu.

Bellez, S., Roussel, H., Dahon, C., Geffrin, J.-M. (2011). A rigorous forest scattering model validation through comparison with indoor bistatic scattering measurements. *Progress in Electromagnetics Research*, 33, 1–19.

Bilicz, S. (2011). Application of design-of-experiment methods and surrogate models in electromagnetic nondestructive evaluation. PhD Thesis, Université Paris Sud, Paris [Online]. Available at: https://tel.archives-ouvertes.fr/ tel-00601753.

Brigot, G., Simard, M., Colin, E., Boulch, A. (2019). Retrieval of forest vertical structure from PolInSAR data by machine learning using LIDAR-derived features. *Remote Sensing*, 11(4), 381.

Brigui, F., Thirion-Lefevre, L., Ginolhac, G., Forster, P. (2013). New SAR algorithm based on orthogonal projections for MMT detection and interference reduction. *IEEE Transactions on Geoscience and Remote Sensing*, 52(7), 3800–3811.

Brigui, F., Ginolhac, G., Thirion-Lefevre, L., Forster, P. (2014). New SAR target imaging algorithm based on oblique projection for clutter reduction. *IEEE Transactions on Aerospace and Electronic Systems*, 50(2), 1118–1137.

Brisco, B., Ahern, F., Hong, S.-H., Wdowinski, S., Murnaghan, K., White, L., Atwood, D.K. (2015). Polarimetric decompositions of temperate wetlands at C-band. *IEEE Journal of Selected Topics in Applied Earth Observations and Remote Sensing*, 8(7), 3585–3594.

Cantalloube, H. and Colin, E. (2008). Assessment of physical limitations of high resolution on targets at X-band from circular SAR experiments. In *7th European Conference on Synthetic Aperture Radar*. VDE, Friedrichshafen.

CCR (2019). Making government science work for Canada and the world, a legacy of excellence at the Canada Centre for Remote Sensing (CCRS). Report, Canadian Government.

Chauhan, N.S., Lang, R.H., Ranson, K.J. (1991). Radar modeling of a boreal forest. *IEEE Transactions on Geoscience and Remote Sensing*, 29(4), 627–638.

Cloude, S.R. and Papathanassiou, K. (2003). Three-stage inversion process for polarimetric SAR interferometry. *IEEE Proceedings-Radar, Sonar and Navigation*, 150(3), 125–134.

Cloude, S.R. and Pottier, E. (1996). A review of target decomposition theorems in radar polarimetry. *IEEE Transactions on Geoscience and Remote Sensing*, 34(2), 498–518.

Cloude, S.R., Corr, D., Williams, M.L. (2004). Target detection beneath foliage using polarimetric synthetic aperture radar interferometry. *Waves in Random Media*, 14, S393–S414.

Colin, E., Titin-Schnaider, C., Thirion, L., Tabbara, W. (2004). Comparison between simulations and interferometric polarimetric SAR P-band data on a pine-trees forest. In *2004 IEEE International Geoscience and Remote Sensing Symposium*, Anchorage.

Colin, E., Titin-Schnaider, C., Tabbara, W. (2005). FOPEN with polarimetric interferometry: Validations with experimental data at P-band. In *Proceedings of the 2nd International Workshop POLINSAR 2005 (ESA SP-586)*. ESRIN, Frascati.

Couderc, O. (2019). Inversion de permittivités d'une structure dièdrique au moyen de la polarimétrie radar. PhD Thesis, Université Paris-Saclay, Paris.

Couderc, O., Thirion-Lefevre, L., Guinvarc'h, R. (2018). Analytical solution for permittivities of a dihedral configuration. *IEEE Antennas and Wireless Propagation Letters*, 17(3), 485–488.

Di Martino, T., Guinvarc'h, R., Thirion-Lefevre, L., Colin, E. (2021). Beets or cotton? Blind extraction of fine agricultural classes using a convolutional autoencoder applied to temporal SAR signatures. *IEEE Transactions on Geoscience and Remote Sensing*, 60, 5212018. doi: 10.1109/TGRS.2021.3100637.

Dubois-Fernandez, P., Cantalloube, H., Vaizan, B., Krieger, G., Horn, R., Wendler, M., Giroux, V. (2006). ONERA-DLR bistatic SAR campaign: Planning, data acquisition, and first analysis of bistatic scattering behaviour of natural and urban targets. *IEEE Proceedings - Radar, Sonar and Navigation*, 153, 214–223(9) [Online]. Available at: https://digital-library.theiet.org/content/journals/10.1049/ip-rsn_20045117.

Durden, S.L., Van Zyl, J.J., Zebker, H.A. (1989). Modeling and observation of the radar polarization signature of forested areas. *IEEE Transactions on Geoscience and Remote Sensing*, 27(3), 290–301.

Everaere, E. (2015). Polarimetry in bistatic configuration for ultra high frequency radar measurements on forest environment. PhD Thesis, École Polytechnique, Palaiseau.

Everaere, E., Colin, E., Thirion-Lefevre, L., Tran, J., Sorin-Cojocaru, C., De Martino, A., Ossikovski, R. (2016). Optical scale polarimetric device for nanotube forest measurement: An opportunity to anticipate bistatic polarimetric SAR images of tree trunk forests at P-band. *IEEE Journal of Selected Topics in Applied Earth Observations and Remote Sensing*, 9(7), 3249–3258.

Freeman, A. and Durden, S.L. (1993). Three-component scattering model to describe polarimetric SAR data. *Radar Polarimetry*, 1748, 213–224.

Ginolhac, G., Thirion-Lefevre, L., Durand, R., Forster, P. (2010). SAR processors based on signal or interference subspace detector. *IEEE Transactions on Aerospace and Electronic Systems*, 46(3), 1006–1020.

Imhoff, M.L. (1995). Radar backscatter and biomass saturation: Ramifications for global biomass inventory. *IEEE Transactions on Geoscience and Remote Sensing*, 33(2), 511–518.

Imhoff, M.L., Story, M., Vermillion, C., Khan, F., Polcyn, F. (1986). Forest canopy characterization and vegetation penetration assessment with space-borne radar. *IEEE Transactions on Geoscience and Remote Sensing*, 4, 535–542.

Israelsson, H., Ulander, L.M., Martin, T., Askne, J.I. (2000). A coherent scattering model to determine forest backscattering in the VHF-band. *IEEE Transactions on Geoscience and Remote Sensing*, 38(1), 238–248.

Jaeger, M. and De Reffye, P. (1992). Basic concepts of computer simulation of plant growth. *Journal of Biosciences*, 17(3), 275–291.

JPL (1984). Jet Propulsion Laboratory annual report.

Karam, M.A. and Fung, A.K. (1988). Electromagnetic scattering from a layer of finite length, randomly oriented, dielectric, circular cylinders over a rough interface with application to vegetation. *International Journal of Remote Sensing*, 9(6), 1109–1134.

Karam, M.A., Fung, A.K., Lang, R.H., Chauhan, N.S. (1992). A microwave scattering model for layered vegetation. *IEEE Transactions on Geoscience and Remote Sensing*, 30(4), 767–784.

Kobayashi, Y., Sarabandi, K., Pierce, L., Dobson, M.C. (2000). An evaluation of the JPL TOPSAR for extracting tree heights. *IEEE Transactions on Geoscience and Remote Sensing*, 38(6), 2446–2454.

Kwoun, O.-I. and Lu, Z. (2009). Multi-temporal RADARSAT-1 and ERS backscattering signatures of coastal wetlands in southeastern Louisiana. *Photogrammetric Engineering & Remote Sensing*, 75(5), 607–617.

Lebarbu, C., Ceotto, E., Cochin, C., Jouade, A., Everaere, E., Trouve, N. (2021). Complete radar simulation chain: Application to maritime patrol surveillance using SAR/ISAR modes. In *EUSAR 2021 – 13th European Conference on Synthetic Aperture Radar*, Leipzig.

Lee, J.-S. and Pottier, E. (2017). *Polarimetric Radar Imaging: From Basics to Applications*. CRC Press, Boca Raton.

Lesturgie, M., Thirion-Lefevre, L., Saillant, S., Dorey, P. (2016). A new technique to characterize foliage attenuation using passive radar in the L-band. *Comptes Rendus Physique*, 17(9), 1003–1017.

Li, H., Li, Q., Wu, G., Chen, J., Liang, S. (2016). The impacts of building orientation on polarimetric orientation angle estimation and model-based decomposition for multilook polarimetric SAR data in urban areas. *IEEE Transactions on Geoscience and Remote Sensing*, 54(9), 5520–5532.

Lin, Y.-C. and Sarabandi, K. (1999). A Monte Carlo coherent scattering model for forest canopies using fractal-generated trees. *IEEE Transactions on Geoscience and Remote Sensing*, 37(1), 440–451.

López-Sánchez, J.M., Esteban-González, H., Baquero-Escudero, M., Fortuny-Guasch, J. (1999). An electromagnetic scattering model for multiple tree trunks above a tilted rough ground plane. *IEEE Transactions on Geoscience and Remote Sensing*, 37(2), 659–667.

López-Sanchez, J.M., Ballester-Berman, J.D., Marquez-Moreno, Y. (2007). Model limitations and parameter-estimation methods for agricultural applications of polarimetric SAR interferometry. *IEEE Transactions on Geoscience and Remote Sensing*, 45(11), 3481–3493.

Maghsoudi, Y., Collins, M.J., Leckie, D.G. (2013). Radarsat-2 polarimetric SAR data for boreal forest classification using SVM and a wrapper feature selector. *IEEE Journal of Selected Topics in Applied Earth Observations and Remote Sensing*, 6(3), 1531–1538.

Martinez, J.-M., Floury, N., Le Toan, T., Beaudoin, A., Hallikainen, M.T., Makynen, M. (2000). Measurements and modeling of vertical backscatter distribution in forest canopy. *IEEE Transactions on Geoscience and Remote Sensing*, 38(2), 710–719.

Massonnet, D., Rossi, M., Carmona, C., Adragna, F., Peltzer, G., Feigl, K., Rabaute, T. (1993). The displacement field of the Landers earthquake mapped by radar interferometry. *Nature*, 364(6433), 138–142.

Melon, P., Martinez, J.-M., Le Toan, T., Ulander, L.M., Beaudoin, A. (2001). On the retrieving of forest stem volume from VHF SAR data: Observation and modeling. *IEEE Transactions on Geoscience and Remote Sensing*, 39(11), 2364–2372.

Nagai, S., Nasahara, K.N., Akitsu, T.K., Saitoh, T.M., Muraoka, H. (2020). Importance of the collection of abundant ground-truth data for accurate detection of spatial and temporal variability of vegetation by satellite remote sensing. *Biogeochemical Cycles: Ecological Drivers and Environmental Impact*, 223–244. doi: 10.1002/9781119413332.ch11.

Nakamura, K., Doi, K., Shibuya, K. (2007). Estimation of seasonal changes in the flow of Shirase Glacier using JERS-1/SAR image correlation. *Polar Science*, 1(2–4), 73–83.

Nguyen, H., Roussel, H., Tabbara, W. (2006). A coherent model of forest scattering and SAR imaging in the VHF and UHF-band. *IEEE Transactions on Geoscience and Remote Sensing*, 44(4), 838–848.

Paillou, N., Thirion-Lefevre, L., Guinvarc'h, R. (2020). Interest of temporal methods over spatial methods in order to detect small targets. In *IGARSS 2020 – 2020 IEEE International Geoscience and Remote Sensing Symposium*, Waikoloa.

Papathanassiou, K.P., Cloude, S.R., Pardini, M., Quiñones, M.J., Hoekman, D., Ferro-Famil, L., Goodenough, D., Chen, H., Tebaldini, S., Neumann, M. et al. (2021). Forest applications. *Polarimetric Synthetic Aperture Radar: Principles and Application*, 59–117.

Proisy, C., Mougin, E., Dufrêne, E., Le Dantec, V. (2000). Monitoring seasonal changes of a mixed temperate forest using ERS SAR observations. *IEEE Transactions on Geoscience and Remote Sensing*, 38(1), 540–552.

Quegan, S., Toan, T.L., Chave, J., Dall, J., Exbrayat, J.-F., Minh, D.H.T., Lomas, M., D'Alessandro, M.M., Paillou, P., Papathanassiou, K. et al. (2019). The European Space Agency BIOMASS mission: Measuring forest above-ground biomass from space. *Remote Sensing of Environment*, 227, 44–60. [Online]. Available at: https://www. sciencedirect.com/science/article/pii/S0034425719301233.

Ranson, K. and Smith, J.A. (1990). Airborne SAR experiment for forest ecosystems research Maine 1989 experiment. In *10th Annual International Symposium on Geoscience and Remote Sensing*, College Park, MD.

Richards, J.A., Sun, G.-Q., Simonett, D.S. (1987). L-band radar backscatter modeling of forest stands. *IEEE Transactions on Geoscience and Remote Sensing*, 4, 487–498.

Romero-Puig, N., Marino, A., Lopez-Sanchez, J.M. (2021). Application of the trace coherence to HH-VV PolInSAR TanDEM-X data for vegetation height estimation. *IEEE Transactions on Geoscience and Remote Sensing*, 60, 4404210. doi: 10.1109/TGRS.2021.3101016.

Ruck, G. (1970). *Radar Cross Section Handbook*, volume 1. Springer, Berlin.

Ruck, G., Barrick, D.E., Stuart, W.D. (1970). *Radar Cross Section Handbook*, volume 2. Springer, Berlin.

Saatchi, S.S. and McDonald, K.C. (1997). Coherent effects in microwave backscattering models for forest canopies. *IEEE Transactions on Geoscience and Remote Sensing*, 35(4), 1032–1044.

Sader, S. (1987). Forest biomass, canopy structure, and species composition relationships with multipolarization L-band synthetic aperture radar data. *Photogrammetric Engineering and Remote Sensing*, 53, 193–202.

Sarabandi, K. and Lin, Y.-C. (2000). Simulation of interferometric SAR response for characterizing the scattering phase center statistics of forest canopies. *IEEE Transactions on Geoscience and Remote Sensing*, 38(1), 115–125.

Staples, G. and Rodrigues, D.F. (2013). Maritime environmental surveillance with RADARSAT-2. In *Anais XVI Simposio Brasileiro de Sensoriamento Remoto-SBSR*, Foz do Iguaçu.

Sun, G., Simonett, D.S., Strahler, A.H. (1991). A radar backscatter model for discontinuous coniferous forests. *IEEE Transactions on Geoscience and Remote Sensing*, 29(4), 639–650.

Thirion-Lefevre, L. and Colin, E. (2005). On the use of a coherent scattering model to determine the origin of artificial signatures of a target hidden in a forest. In *Proceedings of the 2nd International Workshop POLINSAR 2005 (ESA SP-586)*. ESRIN, Frascati.

Thirion-Lefevre, L. and Colin, E. (2007). Investigating attenuation, scattering phase center, and total height using simulated interferometric SAR images of forested areas. *IEEE Transactions on Geoscience and Remote Sensing*, 45(10), 3172–3179.

Thirion-Lefevre, L. and Guinvarc'h, R. (2018). The double Brewster angle effect. *Comptes Rendus Physique*, 19(1/2), 43–53.

Thirion-Lefevre, L., Colin, E., Dahon, C. (2006). Capabilities of a forest coherent scattering model applied to radiometry, interferometry, and polarimetry at P-and L-band. *IEEE Transactions on Geoscience and Remote Sensing*, 44(4), 849–862.

Thirion-Lefevre, L., Bilicz, S., Lambert, M., Gyimothy, S. (2010a). On the use of an "optimal database" for radar forest observation. In *8th European Conference on Synthetic Aperture Radar*, Aachen.

Thirion-Lefevre, L., Colin, E., Dahon, C. (2010b). Bistatic scattering from forest components. Part I: Coherent polarimetric modelling and analysis of simulated results. *Waves in Random and Complex Media*, 20(1), 36–61. doi: 10.1080/17455030903499680.

Thirion-Lefevre, L., Guinvarc'h, R., Colin, E. (2020). The combined effect of orientation angle and material on PolSAR images of urban areas. *Remote Sensing*, 12(10), 1632.

Touzi, R. (2016). Polarimetric target scattering decomposition: A review. In *2016 IEEE International Geoscience and Remote Sensing Symposium (IGARSS)*, Beijing.

Treuhaft, R.N. and Cloude, S.R. (1999). The structure of oriented vegetation from polarimetric interferometry. *IEEE Transactions on Geoscience and Remote Sensing*, 37(5), 2620–2624.

Treuhaft, R.N., Madsen, S.N., Moghaddam, M., van Zyl, J.J. (1996). Vegetation characteristics and underlying topography from interferometric radar. *Radio Science*, 31(6), 1449–1485.

Trouve, N., Colin, E., Fargette, P., De Martino, A. (2011). Influence of geometrical configurations and polarization basis definitions on the analysis of bistatic polarimetric measurements. *IEEE Transactions on Geoscience and Remote Sensing*, 49(6), 2238–2250.

Ulaby, F.T. and Elachi, C. (1990). *Radar Polarimetry for Geoscience Applications*. Norwood, London.

Ulaby, F.T., Sarabandi, K., McDonald, K., Whitt, M., Dobson, M.C. (1990). Michigan microwave canopy scattering model. *International Journal of Remote Sensing*, 11(7), 1223–1253.

Vasko, A., Thirion-Lefèvre, L., Bilicz, S., Champion, I., Lambert, M., Gyimothy, S. (2011). Metamodel-based adaptive use of a coherent polarimetric backscattering simulator for the characterization of forested areas at low frequencies. In *PIERS 2011 – Progress in Electromagnetics Research Symposium, PIERS Proceedings*, Suzhou [Online]. Available at: https://hal.inrae.fr/hal-02749058.

Villard, L., Borderies, P., Le Toan, T., Koleck, T., Albinet, C. (2010). Topography effects on forest radar scattering, consequences on biomass retrieval. In *2010 IEEE International Geoscience and Remote Sensing Symposium*, Honolulu.

Weissgerber, F., Colin, E., Trouvé, N., Nicolas, J.-M. (2016). A temporal estimation of entropy and its comparison with spatial estimations on PolSAR images. *IEEE Journal of Selected Topics in Applied Earth Observations and Remote Sensing*, 9(8), 3809–3820.

Weissgerber, F., Colin, E., Nicolas, J.-M., Trouvé, N. (2017). 3D monitoring of buildings using TerraSAR-X InSAR, DInSAR and PolSAR capacities. *Remote Sensing*, 9(10), 1010.

Williams, M.L., Pottier, E., Ferro-Famil, L., Allain, S., Cloude, S., Hajnsek, I., Papathanassiou, K., Moreira, A., Minchella, A., Desnos, Y. (2007). Forest coherent SAR simulator within PolSARpro: An educational toolbox for PolSAR and PolInSAR data processing. In *Asian Conference on Remote Sensing*, Kuala Lumpur.

Wu, S.-T. and Sader, S.A. (1987). Multipolarization SAR data for surface feature delineation and forest vegetation characterization. *IEEE Transactions on Geoscience and Remote Sensing*, 1, 67–76.

Yamaguchi, Y., Moriyama, T., Ishido, M., Yamada, H. (2005). Four-component scattering model for polarimetric SAR image decomposition. *IEEE Transactions on Geoscience and Remote Sensing*, 43(8), 1699–1706.

Yokoya, N., Yamanoi, K., He, W., Baier, G., Adriano, B., Miura, H., Oishi, S. (2020). Breaking limits of remote sensing by deep learning from simulated data for flood and debris-flow mapping. *IEEE Transactions on Geoscience and Remote Sensing*, 60, 4400115.

Yueh, S.H., Kong, J.A., Jao, J.K., Shin, R.T., Le Toan, T. (1992). Branching model for vegetation. *IEEE Transactions on Geoscience and Remote Sensing*, 30(2), 390–402.

Zebker, H.A., Rosen, P.A., Goldstein, R.M., Gabriel, A., Werner, C.L. (1994). On the derivation of coseismic displacement fields using differential radar interferometry: The Landers earthquake. *Journal of Geophysical Research: Solid Earth*, 99(B10), 19617–19634.

Zhu, X.X., Tuia, D., Mou, L., Xia, G.-S., Zhang, L., Xu, F., Fraundorfer, F. (2017). Deep learning in remote sensing: A comprehensive review and list of resources. *IEEE Geoscience and Remote Sensing Magazine*, 5(4), 8–36.

6

Radiative Transfer Model Inversion and Application to Coastal Observation

Touria BAJJOUK[1], Audrey MINGHELLI[2], Malik CHAMI[3] and Tristan PETIT[4]

[1] *Ifremer, DYNECO, LEBCO, Plouzané, France*
[2] *LIS, CNRS, Université de Toulon, France*
[3] *LATMOS, CNRS, Sorbonne Université, Paris, France*
[4] *Norwegian Polar Institute, Tromsø, Norway*

6.1. Introduction

Being a highly coveted space due to the numerous goods and services they provide in terms of food, recreational services, biodiversity or seascapes, coastal areas are subject to perpetual changes with natural (climate change) and anthropogenic origins, which are directly or indirectly generated by the human activities that these areas foster. These evolutions are reflected by a strong spatial and temporal variability in their bio-optical, morphological and biogeochemical properties, which need to be observed in order to (i) better understand the functioning of these dynamic and complex ecosystems; and (ii) consequently support management measures for maintaining their ecological quality.

For a color version of all the figures in this chapter, see www.iste.co.uk/yan/remotesensing.zip.

Inversion and Data Assimilation in Remote Sensing,

Due to the development of new sensors, optical remote sensing, in the 1990s, became an alternative, making the monitoring of coastal areas on a large scale and at a lower cost possible (Mumby et al. 1999). Nonetheless, the complexity and high spatial and temporal variability of these areas make their study particularly difficult, especially due to the absorption of light radiation by the water column. In fact, the background reflectance amplitude becomes variable, depending on the depth.

The remote sensing research community first alluded, in the mid-1970s, to the potential of satellite data for studies related to signal penetration in water (Sturm 1981; Hick and Scoones 1988). Initially, very simple image processing and enhancement techniques were applied to remote sensing data to obtain estimates of substrate type and/or bathymetry. Then, a number of spectral techniques were developed to determine dedicated biogeophysical properties from oceanographic hyperspectral data (Chang et al. 2004; Fearns et al. 2011). These techniques were based on radiative transfer modeling (RTM). The inversion of these models applied to hyperspectral data led not only to obtaining the specific seabed reflectance and its characterization (Mishra et al. 2007), but also to estimating water constituents (Lorthiois et al. 2012) as well as the bathymetry (Minghelli-Roman et al. 2009).

What is a radiative transfer model? What are the methods used for its inversion? For which applications? By answering these questions in a summarized manner, this chapter will bring new insight about the use of marine radiation for the characterization of coastal areas.

6.2. Principle and treatment method

6.2.1. *Inherent optical water properties*

Solar radiation propagation in the water column is characterized by the interaction of this radiation with the marine constituents encountered during the light path. These marine constituents, namely water molecules, dissolved organic matter, organic (such as phytoplankton) or mineral marine particles, called hydrosols, affect the propagation of light by means of two optical processes: absorption and scattering. Absorption represents a transformation of radiant energy into heat. Therefore, light photons no longer propagate in the medium, but their energy nevertheless remains preserved in the hydrosols that have interacted with them. To quantify the phenomenon of light absorption by hydrosols, the absorption coefficient, denoted by a, is defined as the ratio between the energy flow absorbed by the medium and the incident energy flux for a water thickness of 1 m. The unit of the absorption coefficient is therefore the m^{-1}. It should be noted that the inverse of the absorption coefficient corresponds to the average distance traveled by a photon (in m) before being absorbed by marine constituents (water molecules, marine particles). Scattering characterizes the angular distribution of radiation that has interacted with a marine constituent. In this case, the light photon carries on its propagation but its amount of energy is distributed all around the particle. In particular, a greater amount of energy will be directed in the direction corresponding to the incidence angle of the incoming light (region called "forward scattering"), while lesser energy will be directed in the opposite direction to the incident radiation (region called "backscattering"). We also observe that the energy of the incident photon is preserved after its interaction with the marine constituent. To quantify the scattering phenomenon, the scattering coefficient, denoted by b, is defined as the ratio between the energy flow scattered by the medium and the incident energy flux for a water thickness of 1 m. Its unit is the m^{-1}. The sum of the absorption and scattering coefficient defines the attenuation coefficient, denoted by c.

The coefficients of absorption, scattering and attenuation constitute the inherent optical properties of the medium. These properties depend only on the intrinsic characteristics of the environment under consideration (composition, size and concentration of marine constituents). In particular, they are independent of the amount of incident light or its angular distribution. In contrast, the inherent optical properties are highly dependent on the wavelength (i.e. the color) of the incident light. Finally, it should be emphasized that the inherent optical properties are additive; for example, a medium absorption coefficient is the sum of the absorption coefficients of all marine constituents existing in the environment.

6.2.2. *Apparent optical properties of water*

Apparent optical properties are radiometric quantities that depend on the absorption and scattering coefficients of the medium, but also on the nature of the ambient light field (intensity, angular distribution). The reference apparent optical property is the radiance that represents the radiant energy flux emitted or received by a medium per unit area and for a given direction per unit solid angle (in steradian). Its unit is the $\mathrm{Wm}^{-2}\mathrm{sr}^{-1}$. Other radiometric quantities are frequently used such as irradiance E, which is defined as the energy flux received (or emitted) per unit area. The ratio between the upwelling irradiance coming out of the water column, denoted E_u, and the downwelling irradiance at surface level E_d, defines the reflectance R ($= E_u/E_d$). In the field of satellite remote sensing, what is referred to as "remote-sensing" reflectance (denoted by Rrs) is defined as the ratio between the radiance coming out of the water, L_w, and the downwelling irradiance E_d at sea surface level. This definition was adopted because of the peculiarity of satellite sensors to observe the surface in a given direction; the upwelling radiance is then the radiometric quantity measured by a satellite sensor. These apparent optical properties can be studied to know and quantify the nature of the radiation that propagates in the coupled atmosphere-ocean system. More particularly, marine reflectance reflects the influence of hydrosols on the observed light radiation, including its color. This is why the field of study of the propagation of solar radiation in the water column using radiometric instruments such as satellite sensors is commonly referred to as "ocean color".

6.3. Biophysical model of radiative transfer

The mathematical formalization of the physical phenomenon of radiation propagation in the coupled atmosphere-ocean system (or atmosphere/continental surfaces) is called "radiative transfer modeling". Therefore, the link between the inherent and apparent optical properties in a medium is represented by the radiative transfer equation. In the field of the color of the ocean, radiation reaching the top of the atmosphere, L_{TOA}, can be written with respect to radiance from each medium (ocean and atmosphere) as follows:

$$L_{TOA} = t * L_w + L_{molecules} + L_{aerosols} + L_{molecules-aerosols} + T * L_{GLINT} \quad [6.1]$$

In equation [6.1], L_{TOA} is the radiance at the top of the atmosphere, L_w is the outgoing radiance of water, $L_{molecules}$ is the radiance induced by air molecules, $L_{aerosol}$ is the radiance induced by atmospheric aerosols, $L_{molecules-aerosols}$ is the radiance induced by the "molecules-aerosols" coupling and L_{GLINT} is the radiance induced by the reflection of solar radiation on the sea surface. The factors "t" and "T" represent what is referred to as the "diffuse" and "direct" transmittance of the atmosphere, respectively. It should be pointed out that the inherent optical properties of marine constituents are taken into account in the term L_w. The objective of modeling the radiative transfer is to simulate the radiation that would be captured by a satellite based on knowledge of the inherent optimal properties of the atmosphere-ocean system. Thereby, a radiative transfer model adapted for the field of study of ocean color must be able to quantify all the terms of equation [6.1] from the knowledge of the optical properties of the marine constituents.

Figure 6.1 illustrates the different contributions to radiation captured by a satellite; these are, in particular, the interactions between the incident solar radiation and the different constituents encountered during the light path (such as air molecules, aerosols and hydrosols).

RTM is important for calibrating satellite sensors, interpreting data acquired by these sensors and developing inversion methods for estimating the inherent optical properties of the water column and the bio-optical properties of marine constituents (composition, concentration, size) from satellite radiation. Many radiative transfer models have been developed in the scientific community. Different methods are used for the development of these models, such as the method of successive orders of scattering, the method of discrete ordinates, the finite element method or the Monte Carlo method.

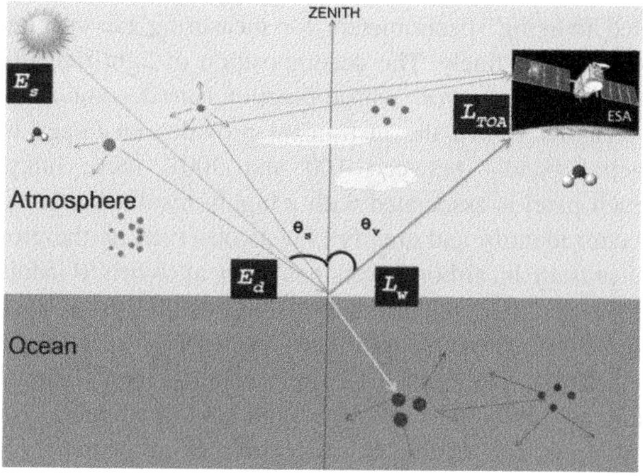

Figure 6.1. *Contributions of radiation reaching a satellite sensor (L_{TOA}) in the ocean–atmosphere system from incident solar irradiance (E_s). E_d is the downwelling solar irradiance at sea surface level; L_w is the outgoing radiance of the water, θ_s and θ_v are the angles of incidence and sight*

An example of a radiative transfer model that is particularly suitable to studying ocean color is the OSOAA model (Ordres Successifs Océan Atmosphère Avancé), which stands for Advanced Successive Orders Ocean Atmosphere (Chami et al. 2015). This model, which solves the radiative transfer equation using the method of successive orders of scattering, is a means to simulate radiation at different depth or altitude levels in the ocean–atmosphere system, while considering complex physical properties such as radiation polarization with a competitive computation time compared to Monte Carlo methods. In addition, it includes a human–machine interface for ease of use. This model is provided in open access to the community by the Centre National dÉtudes Spatiales (CNES, https://github.com/CNES/RadiativeTransferCode-OSOAA).

6.3.1. *Data and preprocessing*

Optical instruments on-board aircraft or satellites measure the radiation they receive from the Earth's surface. Multispectral instruments are equipped with filters that enable radiance measurements for different spectral domains called spectral bands. The number of these bands is small (1 to 10) and they are quite large, which incurs that different spectral information is integrated into a single measurement.

Technological advances in electronics and optics have driven the creation of instruments called imaging spectrometers for measuring the spectral radiances of ground-based objects very finely. The decomposition of light radiation is no longer achieved by filters but by a network whose function is to decompose light for a large number of wavelengths. When the instrument provides an image in a very large number of spectral bands (between 100 and 300), these images are called hyperspectral. Each pixel is associated with a highly resolved reflectance curve that can be used to better identify and qualify objects observed on the ground. This type of instrument is present in airborne versions such as Aviris (Gaddis et al. 1996), Hyspex (Blaaberg et al. 2014), but also in previous versions such as Hico (Keith et al. 2014), currently with Prisma (Anflo and Möllerberg 2009) or with EnMAP (Kaufmann et al. 2006). The French Centre national détudes spatiales (CNES) is currently studying the feasibility of a hyperspectral mission named "Biodiversity" for the purposes of studying coastal aquatic environments (Briottet et al. 2011, 2024).

6.3.1.1. *Atmospheric correction*

The radiance measured at the satellite level (L_{TOA}) differs from the radiance emitted at ground level by the path of the light signal through the atmosphere that modified it. Gases and aerosols existing in the atmosphere absorb and scatter light. To correct for atmospheric effects, assumptions must be made about atmospheric and aerosol models. Firstly, to remove the influence of the observation conditions (acquisition time, day of the year), the TOA radiances (L_{TOA}) are converted into TOA reflectances (ρ_{TOA}):

$$\rho_{TOA} = \frac{\pi L_{TOA}(\lambda)}{dF_0(\lambda)cos(\theta_s)} \qquad [6.2]$$

where θ_s is the solar zenith angle, $F_0(\lambda)$ is the solar irradiance on top of the atmosphere and d is the distance from Earth to the Sun in astronomical units.

The surface reflectance of water ρ_w can then be derived from the surface reflectance by estimating atmospheric reflectance, which is the sum of Rayleigh reflectance (ρ_r) and aerosol reflectance (ρ_a). Models such as Modtran (Berk et al. 1998) and 6S (Vermote et al. 1997) were designed to calculate (ρ_r) as well as atmospheric transmittance (t), which is the proportion of radiated energy transmitted through the atmosphere:

$$\rho_{TOA}(\lambda) = t(\lambda)\rho_w(\lambda) + \rho_r(\lambda) + \rho_a(\lambda) \qquad [6.3]$$

where ρ_r is the reflectance due to Rayleigh scattering and t is the atmospheric transmittance. The reflectance due to aerosol scattering (ρ_a) can be estimated in the image by the "black pixel" method, which assumes that water absorbs all solar radiation in the near-infrared spectral range. The measured signal, after subtraction of Rayleigh reflectance, is then due only to the presence of aerosols. However, in the presence of high concentrations of suspended particulate matter, this assumption is no longer verified. Aerosol reflectance can then be estimated using the spectral bands in the mid-infrared. Software such as Atcor (Richter and Schläpfer 2012), Flaash (Matthew et al. 2002) or Tafkaa (Gao et al. 2006) can be used to enable the correction of atmospheric transfer images.

6.3.1.2. *Correction of surface effects*

The estimation of the contribution of sun reflection onto the sea surface, called *sunglint*, in color images of water at 1 km resolution is based on the Cox and Munk model (Cox and Munk 1954). This model estimates the component of *sunglint* from wind speed and direction (Wang and Bailey 2001; Zhang and Wang 2010). Nonetheless, the wave slope statistics parameterized by the model are not suitable for images with high spatial resolution (namely for a resolution <100 m) (Kay et al. 2009). Therefore, most of the *sunglint* correction methods developed for high-spatial resolution imagery are based on estimating the sun signal without any a priori assumption about the actual state of the sea. The first algorithms developed were designed to utilize the signal measured in NIR (near infrared) where the contribution of radiation from water is assumed to be negligible (Hochberg et al. 2003; Philpot 2007). Improvements to these methods, which were mainly developed for hyperspectral imaging, made it possible to take into account the significant radiation of water output in the NIR by (i) employing the characteristics of peak oxygen absorption around 760 nm (Kutser et al. 2009); (ii) the spectral characteristics of the water column according to a pixel-to-pixel approach (Goodman et al. 2008); or (iii) statistics on a set of light and dark pixels within the image (Hedley et al. 2005; Lyzenga et al. 2006; Martin et al. 2016). These methods are based on radiance regression analyses in the visible region with those measured in the near infrared. Regressions are performed for a given *region of interest* (ROI) of the image. It is assumed in the ROI that (i) glinted and glint-free pixels are present; and (ii) marine radiation is spatially homogeneous. The first condition might be difficult to satisfy when the image contains an almost constant level of reflection in each pixel, which can occur in the presence of waves smaller than the pixel size. Selecting the ROI is therefore a critical step in these methods. To overcome this limitation, Harmel et al. (2018) proposed an approach based on the explicit formulation of solar radiation with particular attention paid to the spectral behavior of this radiation from the mid-infrared to the visible wavelengths.

6.3.2. *Estimation and inversion methods*

Since the beginning of the era of ocean color remote sensing, the variables that are estimated are water composition and bathymetry. Seabed cover mapping has also been performed in shallow-water areas. The analysis of the inverse problem, as well as the interpretation of ocean color in terms of optically active constituents, can be carried out by following three approaches: empirical, radiative transfer (with precomputed correspondence tables) or by inversion of semi-analytical models.

The first approach, called empirical, is entirely based on simple relations established between the remote sensing reflectance (R_{rs}) and the parameters to be estimated Morel and Gordon (1980). The different versions of the empirical algorithm developed by O'Reilly et al. (2000) mainly for the use of CZCS, Modis and SeaWiFS sensors are based on the simple principle of establishing the best statistical link between simultaneous measurements of chlorophyll a (chl) concentrations and reflectance Rrs (ρ_w/π). In order to estimate suspended particulate matter concentrations (SPM), a generic algorithm can be employed with a single spectral band in the 520–700 nm range, thus with many sensors subject to sufficient sensitivity to water color (Nechad et al. 2010). Concerning colored dissolved organic matter (CDOM), Kutser et al. (2005) proposed an empirical model for estimating the absorption coefficient of CDOM at 420 nm. This approach is rather dedicated to multispectral images (<10 bands).

The second approach for estimating water column parameters, bathymetry or seabed type uses the direct radiative transfer model presented in section 6.3 to simulate reflectance for a large number of model input parameters (concentration, background depth, background type, viewing and solar angles) and thus create precomputed tables called *Look up Table* (LUT), which are tables relating the surface reflectances to the input parameters of the model. Therefore, for a surface reflectance measured by the sensor, a search through the lookup table is performed to obtain the corresponding input parameters (Hedley et al. 2009). This approach can be used with both multispectral images (<50 bands) and hyperspectral images (>50 bands).

Finally, the third approach, known as semi-analytical, consists of minimizing a distance between the measurement of surface reflectance by a sensor and a simplified radiative transfer model. In 1998, Lee et al. (1998) proposed a simplified semi-analytic radiative transfer model whose main advantage was to be invertible. Lee's work paved the way for estimating seabed and water column parameters by inverting a radiative transfer model based on a measured reflectance spectrum (ρ_s). The model introduced in Lee et al. (1998) described the reflectance below the water surface:

$$\rho_s(\lambda) = \rho_w(\lambda)(1 - e^{-2k(\lambda)z}) + \rho_b(\lambda)e^{-2k(\lambda)z} \quad [6.4]$$

where ρ_w is the reflectance of an infinite water column, k is the diffuse attenuation coefficient, z is the depth, and ρ_b is the reflectance of the seabed. As an output of the inversion, the input parameters that minimize the distance (chl, SPM, CDOM, z and the seabed abundance) are obtained.

The minimization between the model and the measurement can be achieved using the method of least squares (for example, with the Matlab lsqcurvefit function). Since there are several solutions to inversion (ill-posed problem), constraints are added by setting minimum and maximum bounds for each parameter, which leads to obtaining realistic solutions. This approach is rather dedicated to hyperspectral images that possess a large number of spectral bands (>50 bands).

Several variants have spun off from this inversion. The HOPE method (Lee et al. 1999) considered only one seabed spectrum modulated in amplitude by a coefficient. The Bruce (Klonowski et al. 2007), Bomber (Giardino et al. 2012), Sambucca (Brando et al. 2009) and WASI methods (Gege 2004) considered a linear combination of different types of seabeds (abundance with a sum of 1 or not). Filippi and Kubota (2008) took into account the spatial correlation of adjacent pixels to reduce errors (regularization parameters). Finally, statistical methods such as maximum likelihood made estimates more robust but induced a loss of spatial resolution (Jay and Guillaume 2011).

6.3.3. *Validation methods for inversion products*

Maps of phytoplankton, CDOM and non-algal SPM concentrations, bathymetry as well as the abundance of the main objects that make up the seabed can be obtained by inversion algorithms. In order to validate the estimated quantities, it is important to have access to additional spatialized data, acquired at the study sites simultaneously with the images. These data can be categorized into two types: single-point field observations or space observations from remote sensing (Figure 6.2).

6.3.3.1. *In situ sampling and observations*

The validation of satellite-estimated marine reflectances after the atmospheric correction step is carried out using in situ measurements performed during oceanographic surveys at sea. Radiometric measurements are achieved using a portable spectroradiometer. Associated with a spectralon for calibration, this equipment is employed for acquiring spectral reflectance signatures of different targets of interest. These measurements can be made above the surface, just below sea surface or even at the bottom. However, in situ measurements of these marine reflectances remain a challenge, given the difficult conditions for manipulating instruments for measuring optical properties on board ships. Measurement protocols have been proposed to homogenize marine reflectance data used for validation (IOCCG, 2019). In addition, a radiometer instrument array, called AERONET-OC, was developed to routinely measure radiances coming out of water from fixed coastal sites (Zibordi et al. 2009). Uncertainties in reflectance measurements were also documented in Gergely and Zibordi (2014).

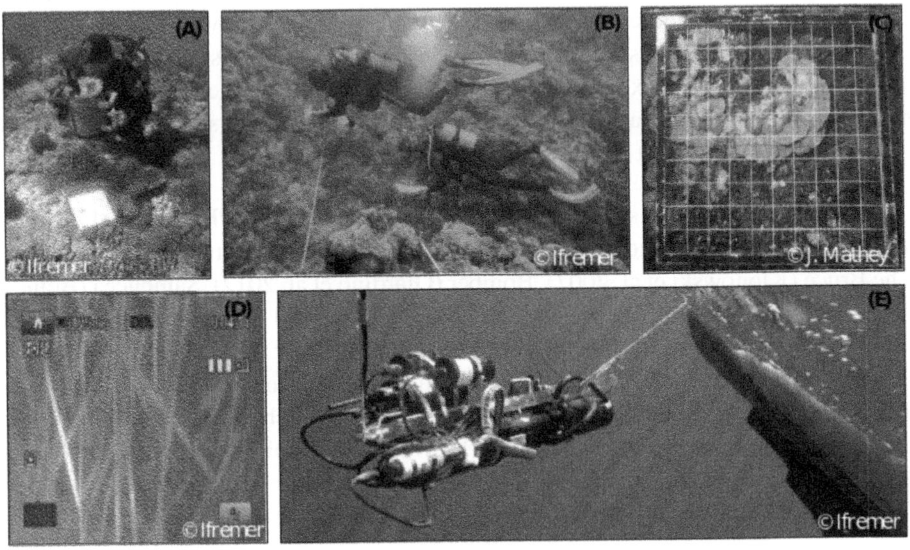

Figure 6.2. *In situ observational data acquisition methods for calibration and validation: reflectance spectra measurements (a), identification and estimation of abundance of seabed types by transect (b), quadrat (c) as well as underwater video (d) and towed optical cluster in open environment (e)*

With regard to the parameters of optical properties of water, their values derived from satellite sensors can be validated by measurements of the chlorophyll concentration carried out in situ: (i) either by filtering water samples, freezing filters and chromatographic analysis by HPLC in the laboratory (*high-performance liquid chromatography* or less commonly *high-pressure liquid chromatography*); (ii) or by a fluorometric probe. The suspended particulate matter concentration is measured from water samples taken and by difference in filter weight after filtration and drying. The 440 nm absorption of dissolved organic matter or CDOM (*colored dissolved organic matter*) is obtained by taking water samples and by measuring absorbance in the laboratory using a spectrometer equipped with a double monochromator beam.

For the validation of the presence/absence and abundance of bottom types, the techniques of measurement in situ are varied and must be adapted to the configuration of the study area. At low tide, field measurements can be carried out using a quadrats to estimate abundance (Figure 6.2). In subtidal areas, these operations are carried out by autonomous divers, by underwater videos acquired by a light-towed system or by underwater drones, preferably using video transmission on the surface. These data are geolocalized using a preferably differential *global positioning system* for improved localization accuracy.

6.3.3.2. *Remote sensing data*

Other data types than in situ observations, based on remote sensing techniques (Figure 6.3), can be employed for the validation of products resulting from the inversion of the radiative transfer model.

Lidar (*LIght Detection And Ranging*), an active laser measurement system, is typically utilized for bathymetry validation (Populus 2002). Since it concerns types of seabeds, acoustic data can also be considered for their validation. The use of lateral sonar has indeed gradually developed in sedimentology, then for mapping biocenoses such as posidonia meadows (Piazzi et al. 2000) or crepidula banks (Sauriau et al. 1998). Moreover, the single-beam sonar is also capable of measuring both water height and detecting targets in the water column. Widely employed in fisheries for the recognition of fish schools and stock assessments, its use can also be extended to the characterization of seabeds (Jordan et al. 2005). Devices known as *split-beam* sounders are also utilized for calibrating the level of the echoes that are received and performing quantitative measurements.

It should be noted that bathymetry validation, resulting from inversion with a Lidar, incurs two major limitations in areas where the bathymetric gradient is large (Petit et al. 2017): (1) the accuracy of georeferencing is not sufficient to reliably compare numerical terrain models (NTMs) created by Lidar and multi- or hyperspectral images; (2) depending on the Lidar technology that is used, the spatial density of Lidar probes may not be sufficient for detecting sudden variations in bathymetry in very rugged environments visible with airborne images of fine spatial resolution. In this case, it is recommended that only areas with low bathymetric gradient should be selected to validate the performance of the estimate.

Concerning the validation of seabed abundances, there are, again, a large number of existing constraints related to the low accuracy of the georeferencing of the ground truth point carried out by the operator, which prevent performing, from ground truths on the nature of the seabeds and the abundance of the habitats therein, a quantitative validation, and especially when addressing highly heterogeneous subtidal areas.

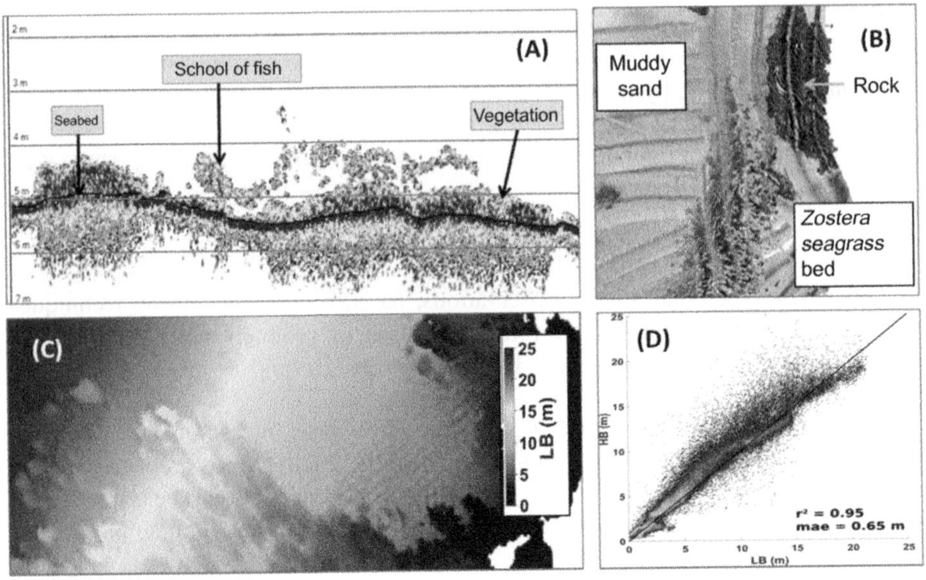

Figure 6.3. *Examples of data used for validating products originating from multi and hyperspectral image inversions*

COMMENT ON FIGURE 6.3.– *(a) Echogram obtained with a single-beam sonar (Bajjouk et al. 2019). (b) Sonar mosaic extract from the bay of Camaret, Finistère, France (Pluquet and Ehrhold 2009). (c) Lidar bathymetry. (d) Comparison with bathymetry ensuing from the applied inversion of the radiative transfer model for hyperspectral images (Petit et al. 2017).*

6.3.4. *Uncertainty about inversion-estimated parameters*

The objective of this section is to concisely present the main specificities of the uncertainty in coastal water color. For a deeper insight into this notion, the reader should refer to Giardino et al. (2019) as well as, more broadly, to the IOCCG report (2019).

The uncertainty in parameter estimation for the water column and/or the seabed depends mainly on (i) the intrinsic quality of the spectral radiance measurement achieved by the sensor; (ii) the optical complexity of the environment of interest; and (iii) the characteristics of the algorithmic processing carried out.

Measurement uncertainty includes a deterministic part related to the inaccuracy of sensor calibration (of the order of a few percent of the measured value) as well as a stochastic part that mainly comprises photonic noise in the visible/near infrared spectral region and reading noise. A detailed description of these phenomena is given in Janesick (2007) for CCD and CMOS technologies, which are those used in water color.

The concept of signal-to-noise ratio (SNR) is fundamental to quantifying the quality of measurement. In the context of remote sensing, SNR is defined as the ratio between measured radiance and standard deviation of noise. This quantity therefore depends not only on the noise characteristics of the sensor, but also on the number of photons responsible for the spectral radiance value read for an element of the data hypercube. For a given scene, this number depends mainly of spatial and spectral resolutions. In practice, for a constant distance between the sensor and the water surface, the SNR decreases with the improvement of the fineness of the spatial and/or spectral resolutions and a trade-off therefore has necessarily to be found between these two parameters. In the satellite field, for example, an SNR of 1,000 can be found for the OLCI/Sentinel-3 sensor (with spatial resolution of 300 m, 21 spectral bands), while the SNR falls to a value of 200 for the HICO hyperspectral sensor (with a spatial resolution of 90 m, 124 spectral bands).

The SNR is a concept mainly used from the point of view of the hyperspectral system and therefore independently of the targeted application. Nevertheless, in the context of water color, the photons carrying information are only those emerging from the water column. According to simulations by Mobley et al. (2011) (https://www.oceanopticsbook.info/) (satellite context, open ocean water), this "effective" signal represents in the visible/NIR range only 0–10% of the signal measured, depending on the wavelength. If an "effective SNR" was to be calculated for the water color theme, taking into account only the effective signal, very low values would then come across. This thus explains the technological challenge represented by the transition from airborne to satellite aspects for the color of coastal water. As a matter of fact, the latter requires the preservation of both fine spatial resolutions (strong spatial dynamics of the phenomena under study) and spectral resolutions (optical complexity of the coastal environment). Recent technical developments have contributed to us currently considering satellite sensors such as EnMAP (Guanter et al. 2015) with specifications of an SNR of 400–495 nm for a spatial resolution (respectively, spectral) of 30 m (respectively, 6.5/10 nm in the visible/NIR/SWIR range for a total of 242 bands).

The high optical complexity of the coastal environment is one of the main sources of uncertainty. It is indeed in these waters that the greatest diversity of optically active elements (phytoplankton, detritus, CDOM, mineral particles) can be found, with very often contributions on total absorption and backscatter of the same order of magnitude. However, the spectral coupling between the different optical properties

(IOPs) is very significant in the visible range, which necessarily makes their respective estimation difficult when no simplifying assumption can no longer be made (e.g. presence of CDOM correlated with the concentration of phytoplankton and negligible concentration of mineral particles). In addition, the concentrations of backscattering materials may be high enough to generate a significant upwelling marine radiance in the NIR and thus an ocean–atmosphere coupling in this spectral range. In this case, it is either necessary to find complex algorithmic solutions for estimating the level and spectral slope of aerosol-related scattering, or to opt for a sensor possessing spectral bands in mid-infrared regions (SWIR) presenting low absorption by water molecules in gaseous form (Moses et al. 2017).

Algorithmic processing utilized in coastal environments are often developed for specific environmental conditions (phytoplankton type cases, non-algal particle size, seabed type and IOP vertical distribution). Their application outside their optimal operating point can induce significant errors in the associated estimation of parameters. In addition, an important distinction must be made between algorithms using a limited number of spectral bands available (mainly empirical algorithms) and those making use of the full spectrum. Regarding empirical algorithms, in the case-I waters (i.e. open ocean), the chlorophyll concentration can be estimated with an error of 10–50% (Campbell et al. 2011). In the coastal zone, the error is 10–70% in the concentration of chl-a (Gohin et al. 2002), 10–30% in the SM (Nechad et al. 2010) and 10–70% on the CDOM (Zibordi et al. 2013). For model inversion methods, the depth error by the Hope method is 5–10% between 2 and 25 m (simulation), 3–7% between 5 and 12 m for the Bruce method. Dekker et al. (2011) compared state-of-the-art HOPE, Bruce, Sambucat, Smlut, Allut methods, which all give an error between 10–30% from 2 to 15 m deep. The maximum likelihood method gives a relative error in the depth between 5–20% from 5 to 12 m (Jay and Guillaume 2011).

6.4. Examples of applications in coastal areas

6.4.1. *SPM/CHL estimate*

6.4.1.1. *Context and issues*

Although coastal waters account for less than 10% of the global ocean, they harbor nearly 50% of oceanic biological productivity. It is therefore important to be able to detect phytoplankton biomass on a global scale using space remote sensing. In addition, these areas are strongly affected by continents or by exchanges with the atmosphere. The result is a complex mixture of marine constituents that impact the quality of bio-optical parameters estimated by satellite. For example, mineral-like particles found in coastal areas originate from different sources such as river inputs, wind inputs, erosion or still resuspension from the seabed.

6.4.1.2. *Product development*

The satellite-based estimation of phytoplankton biomass is performed through the estimation of the concentration of chlorophyll, a pigment (in $mg.m^{-3}$) which is the only pigment present in all phytoplankton species and, consequently, a robust marker of biomass. This pigment has the property of absorbing radiation in the blue range (typically 440 nm) and less in the red (675 nm), which implies a greenish water color when the concentration of phytoplankton biomass is high. The estimation of the suspended particulate matter concentration, denoted by SPM in (mg/L or g/m^3), is achieved by utilizing marine reflectance at wavelengths close to the red (typically 620 nm) or in the near-infrared range (above 800 nm) for waters very rich in SPM. The principle of reflectance inversion methods for obtaining bio-optical parameters (chlorophyll a and SPM) was described in section 1.4.

6.4.1.3. *Use cases*

A space remote sensing use case of ocean color in coastal areas concerns the observation of phytoplankton algal blooms, in particular, harmful (or toxic) algae which are of important socio-economic interest. An illustration thereof is given in Figure 6.4, which shows the remote sensing of harmful phytoplankton species, namely Nodularia spumigena, in the Baltic Sea using the Modis satellite sensor (NASA). This type of observation is important for estimating the quality of waters and marine ecosystems with implications for tourism and industrial activities such as fishing. The information collected by satellite is integrated into observation and modeling systems in order to predict the initiation, extension and duration of these blooms.

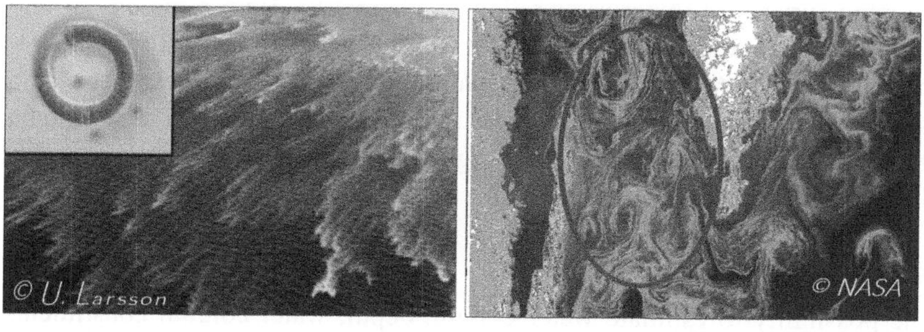

a) b)

Figure 6.4. *Detection of harmful algal bloom Nodularia spumigena in the Baltic Sea observed in situ (a) and (b) by the Modis ocean color sensor (NASA)*

Another remote sensing use case concerns the monitoring of the spatiotemporal extension of the SPM concentration observed by the European satellite sensor Meris (ESA) in the Rhone plume (France) (Figure 6.5). The SPM concentrations provided by this type of observation can be integrated into a dynamic hydro-sedimentary circulation model for estimating material flows at the continent–ocean interface. Satellite SPM monitoring also has implications for coastal water pollution.

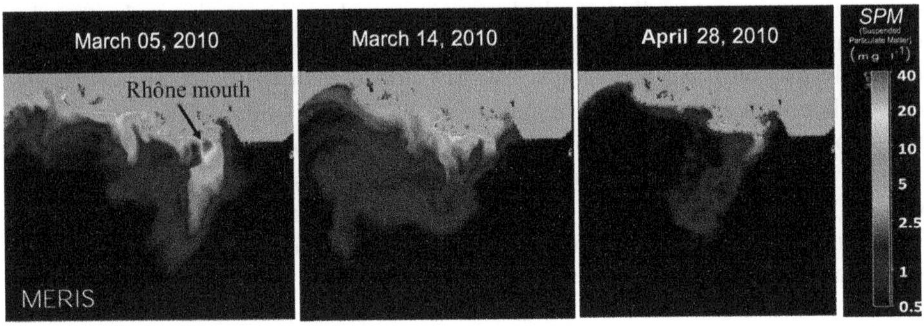

Figure 6.5. *Spatiotemporal monitoring of the Rhône river plume (France) during springtime 2010 (Lorthiois et al. 2012)*

6.4.2. *Bathymetry estimation*

6.4.2.1. *Context and challenges*

In order to explore the seabed, oceanographic ships and underwater vehicles are equipped with multibeam sounders. These acoustic systems are used to obtain topographic surveys of underwater reliefs (bathymetry). With a maximum angle of view of 150°, the swath varies from 74 m (with a depth of 10 m) to 7.4 km (with a depth of 1,000 m). At shallow depths, ships must make many transects (or radial crossings) to cover the coastal strip. Echosounders are therefore not suitable for coastal areas. Bathymetric estimation using optical remote sensing then appears to be an ideal solution.

6.4.2.2. *Product development*

The possibility to estimate water column depth, water column constituents and benthic cover types over large areas and/or remote locations from remote sensing data began in the 2000s, with the first empirical approaches (Lyzenga 1981; Clark et al. 1987; Philpot 1989) that employed only two spectral bands by separating bathymetry information and seabed reflectance information. The disadvantage of this method was that it assumed that the water quality is homogeneous throughout the area, which is not always the case. Lookup table (LUT)-based methods or semi-analytical (SA) models have many advantages for mapping shallow water environments, provided that the image was corrected for atmospheric and water surface effects.

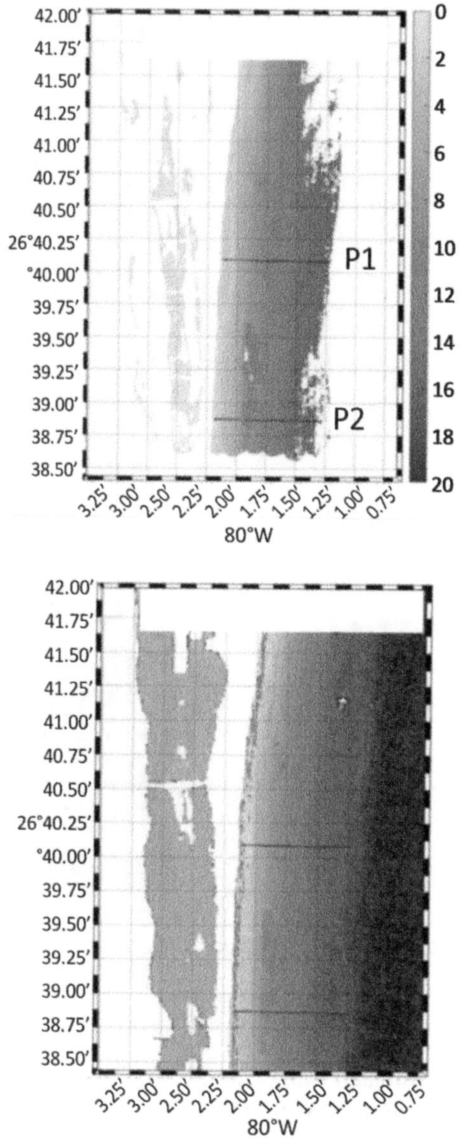

Figure 6.6A. *(a) Airborne Lidar bathymetry at West Palm Beach. (b) Satellite bathymetry*

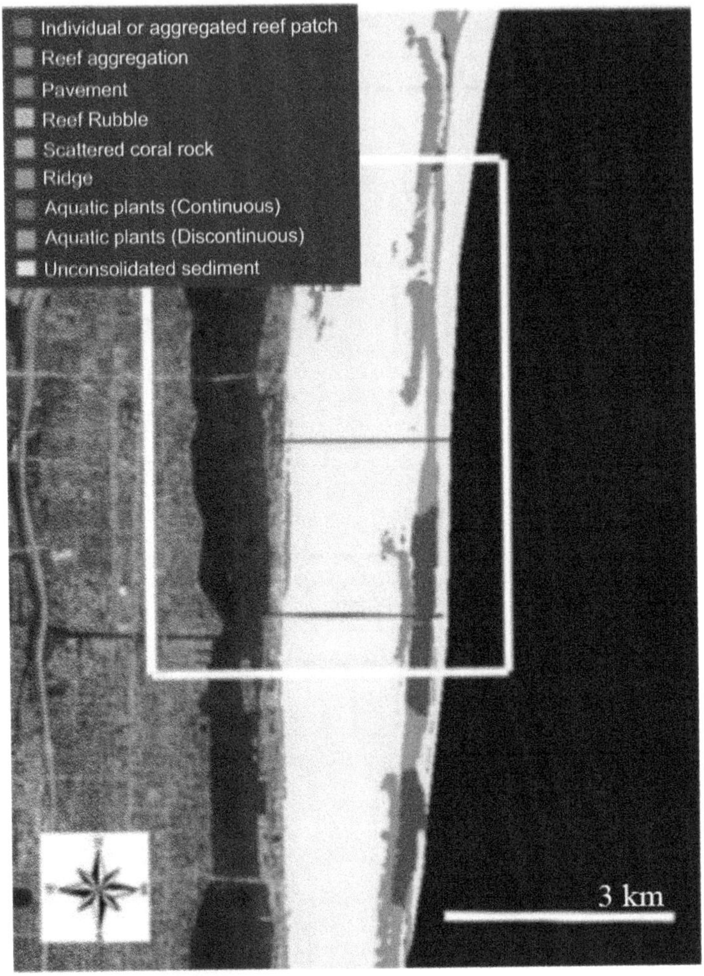

Figure 6.6B. *(c) RGB composite color (Caballero and Stumpf 2019)*

6.4.2.3. *Use cases*

Bathymetry can be estimated using empirical approaches with satellite images of high spatial resolution such as MSI/S2 (Caballero and Stumpf 2019; Figure 6.6B), OLI/Landsat (Jagalingam et al. 2015) and even SPOT (Lafon et al. 2002).

LUT-based approaches or radiative model inversion can be used to estimate bathymetry with aerial or satellite hyperspectral images such as Hyspex, Phills or Casi (Dekker et al. 2011), Chris/Proba or Hyperion (Minghelli-Roman et al. 2009) (Figure 6.7).

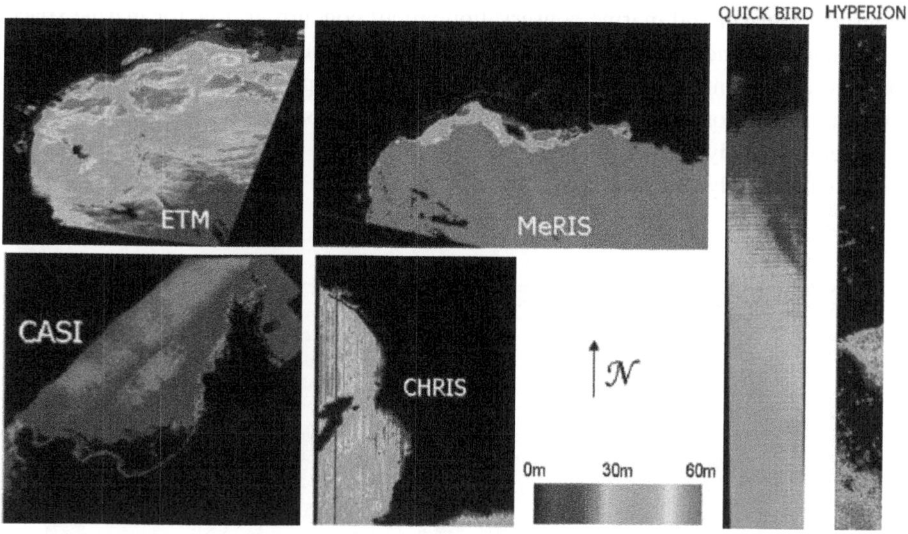

Figure 6.7. *Bathymetric maps obtained from ETM/Landsat, Casi, Meris, Chris/Proba, Quickbird and Hyperion images (Minghelli-Roman et al. 2009)*

To estimate bathymetry, a new method has recently emerged that makes use of the acquisition time delay between panchromatic and multispectral images of Spot or Pléiades images, from which the wave movement speed can be known. Once this velocity (c) and wavelength are estimated in the images, the depth (z) can be deduced by the following relation (Almar et al. 2019):

$$z = \frac{\lambda}{2\pi} tanh^{-1} \frac{2\pi c^2}{g\lambda} \qquad [6.5]$$

with g the acceleration of gravity. The advantage of this method is that it can be applied to turbid waters, but the presence of swell is necessary to be able to retrieve information about the depth. Almar et al. (2019) applied it to Pléiades images (Figure 6.8) with a root mean square error (RMSE) = 1.4 m up to 35 m seabed.

6.4.3. *Spatial characterization of the seabed*

6.4.3.1. *Context and issues*

The way that coastal areas have been used has considerably changed in recent decades. Fishing, aquaculture, shipping, wind farms, the extraction of building materials, tourism and marine conservation are all concurrent uses for marine space. With ongoing concerns about the quality of water bodies and the significant loss of some compartments of marine biodiversity, theme maps of benthic habitats (Nieto 2013) make it possible to provide a fundamental baseline for characterizing the spatiotemporal dynamics of marine systems and for regular monitoring of ecologically highly sensitive areas.

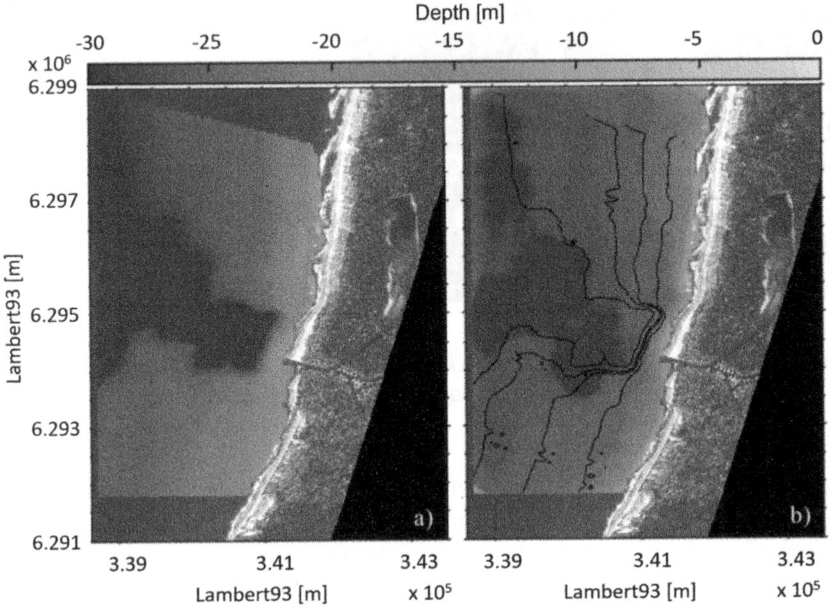

Figure 6.8. *Regional Bathymetry (a) from an echosounder and (b) from Pleiades (Almar et al. 2019)*

The distribution of habitats is also one of the most used pieces of information to meet the recommendations of the Convention on Biological Diversity (United Nations 1992) as well as the requirements of the various European conventions and directives (Water Framework Directive (WFD); Marine Strategy Framework Directive (MSFD) and Habitats Directive for Fauna and Flora (HDFF)).

6.4.3.2. *Product development*

Based on a simplified version of the radiative transfer model, the use of empirical methods for characterizing the seafloor has its origins in the work of Brown et al. (1971) and has since been utilized several times (Lyzenga 1981; Philpot 1989; Mumby et al. 1999; Mishra et al. 2007; Petit et al. 2017). The limitations of empirical methods have led part of the scientific community to rather consider more sophisticated methods for explicitly estimating, in addition to the water height and IOPs of the water column, the abundances of benthic objects. A first family of model inversion methods, called optimization-based methods, has a goal which is to iteratively search for the parameterization of the direct model. This type of method requires the choice of an optimization algorithm, an initial solution, a cost function and the constraints imposed so that the solutions are physically realistic. Regarding background abundances, the physical properties of the spectral mixing phenomenon have led to the development of two complementary constraints widely used by the community (Bioucas-Dias et al. 2012). These are the constraint of non-negativity of abundances and that of sum-to-one. Petit et al. (2017) tested several inversion configurations in coral environments. For most of these configurations, the accuracy of the estimates was better for shallow depths and for lighter seabeds. Despite the fact that seabed types could be detected up to 15–20 m, their thorough discrimination could only be achieved with relatively good accuracy up to 3 m depth. This decrease in discernability with increasing seabed depth was also observed by Goodman and Ustin (2007). The authors showed a close correlation with in situ data for depths between 0 and 3 m.

6.4.3.3. *Use cases*

Most of the remote sensing studies conducted in optically shallow waters aimed to map both the distribution of a single benthic habitat such as seagrass beds (Fornes et al. 2006; Casal et al. 2011) or different benthic habitats coexisting in the same study area (Belluco et al. 2006; Kutser et al. 2006; Bertels et al. 2008; Vahtmäe et al. 2020).

It was possible to distinguish between bare sediments (sand, gravel, limestone, etc.) and benthic vegetation using a minimum distance classification algorithm applied to images in which the water effect was corrected (Vahtmäe et al. 2020, Figure 9). A model inversion method was also evaluated using sensitivity analysis and hyperspectral data to characterize the leaf area index (LAI) of seagrass beds in Florida Bay (Hedley et al. 2017, Figure 6.9). The ability to estimate the LAI was comparable to previously published methods and the derived values per image area properly corresponded to the data measured in situ in terms of mean LAI and standard deviation, despite difficulties in spatial alignment of the in situ data with imagery.

In addition, the spatiotemporal dynamics of the reef platform of Saint-Gilles (Réunion Island) was also studied by inversion of airborne hyperspecral images. Diachronic analysis led to mapping changes in coral health condition between 2009 and 2015 (Figure 6.10), the rate of live coral being as a matter of fact considered the most effective indicator for large-scale application (Holden and LeDrew 2001; Huang et al. 2018). Bajjouk et al. (2019) thus proposed a proxy index of coral health condition, the HCAI (*Hyperspectral Coral to Algae Index*), by estimating the abundance of the main benthic units by unmixing. The authors were able to conclude about the interest of hyperspectral imagery inversion as a useful management tool for facilitating the identification of changes in areas of high conservation challenges, although the health of the coral reef ecosystem can be assessed by a variety of other methods.

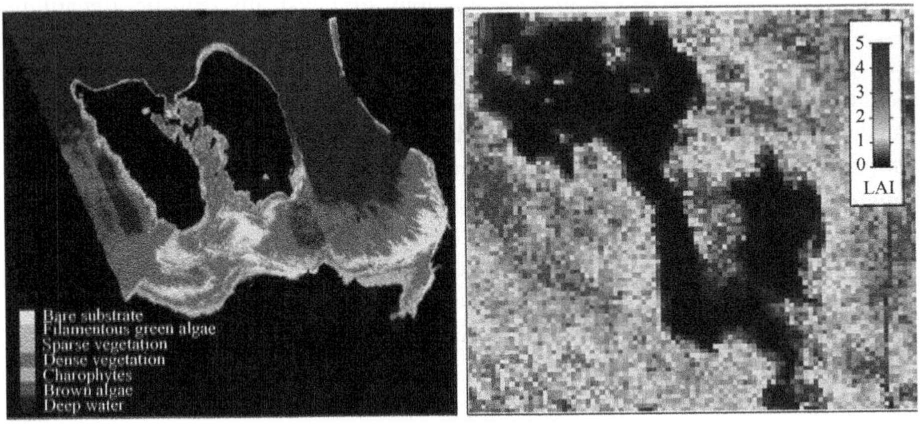

a) b)

Figure 6.9. *a) Map of benthic habitats of the Pakri marine area obtained by supervised classification of airborne imagery CASI (based on Vahtmäe et al. (2020)) and (b) mapping of the leaf area index (LAI) of a Thalassia-dominated Florida Bay seagrass from a Prism hyperspectral image (modified from Hedley et al. (2017))*

6.5. Conclusions and outlook

In coastal areas, several methodologies for taking into account the optical signal of water column effects have emerged over the past 30 years. There is, however, currently no existing consensus regarding the selection of the most efficient method. Indeed, the literature attests to the diversity of studies implementing radiative transfer models, but it should be noted that, despite the progress observed to make inversion methods more robust and efficient, uncertainties associated with the estimated parameters still cannot be avoided. These relate to various factors such as (1) the quality of hyperspectral data (low signal-to-noise ratio and radiometric preprocessing residues); (2) the quality of atmospheric corrections; and (3) the validity of the input parameters of the direct model (IOPs values, limitation of the number of pure targets in the seabed reflectance model).

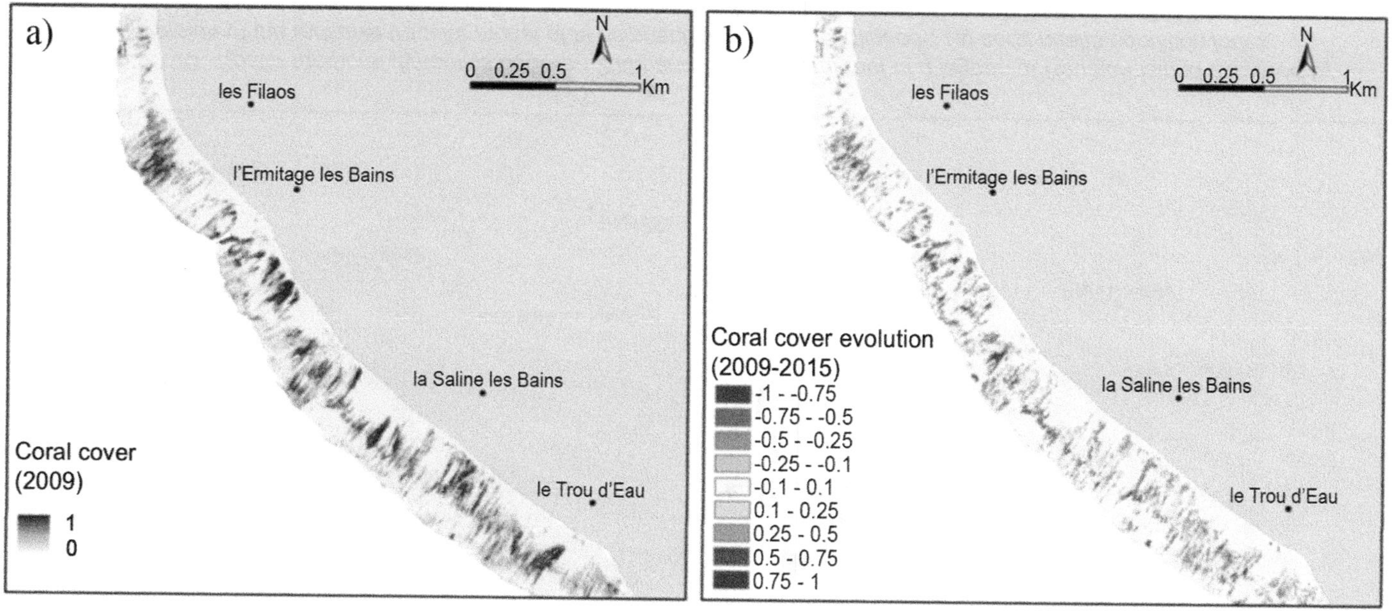

Figure 6.10A. *Spatiotemporal dynamics between 2009 and 2015 of coral reef of a lagoon of Réunion Island achieved by inversion of the radiative transfer model of hyperspectral imaging: (a) coral recovery rate; (b) its change (from Bajjouk et al. (2019))*

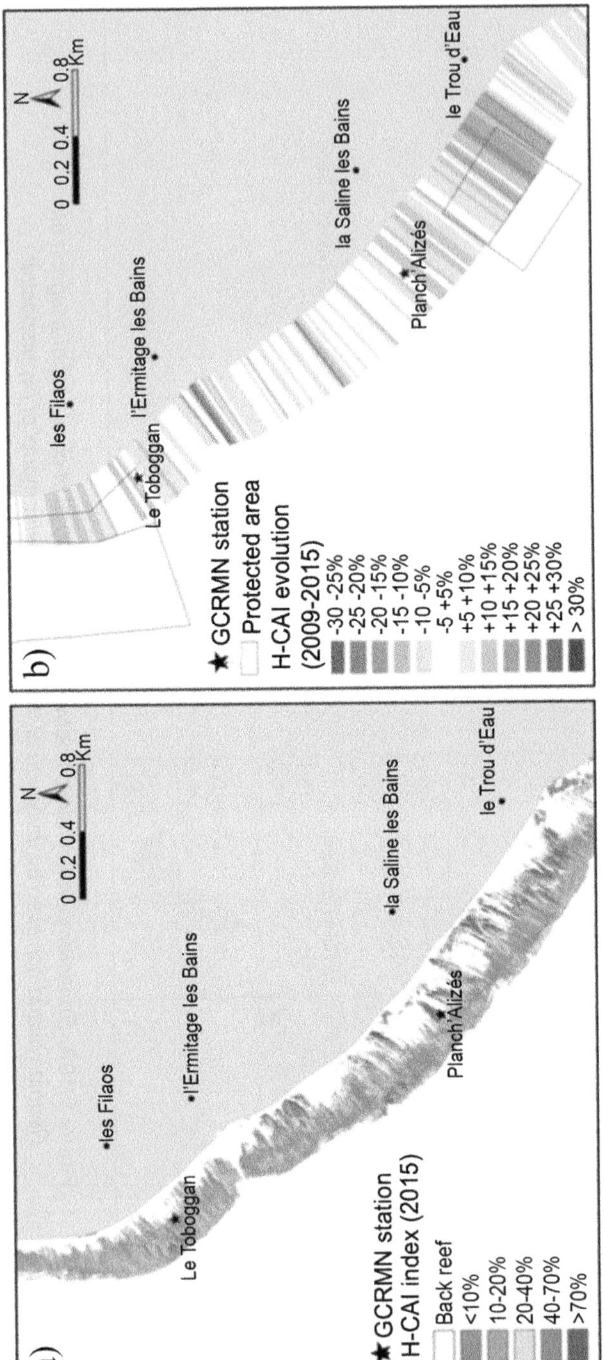

Figure 6.10B. *Spatio-temporal dynamics between 2009 and 2015 of coral reef of a lagoon of Réunion Island achieved by inversion of the radiative transfer model of hyperspectral imaging (continued): (a) coral health condition index HCAI and (b) its aggregation into 20 m lines perpendicular to the coast (from Bajjouk et al. (2019))*

In terms of the validation of inversion products, considerable progress has been made in recent decades with, in particular, the development and operation of high-performance in situ measuring instruments (radiometers) and the development of rigorous sampling protocols for seawater sampling and bio-optical measurements. Nonetheless, the practical considerations for collecting and aligning in situ data with satellite and airborne imagery (i.e. the accurate co-localization of data) remain a major limiting factor in demonstrating the capability of the methods being used. This aspect of experimental design requires further development of protocols in order to better match imaging results with in situ observations and thus allow for quantitative and statistically robust estimation of the retrieved geo-biophysical parameters and their associated uncertainties.

Independently of the limitations identified and the new developments to come for the inversion of multi- and hyperspectral data, the technique of multi- and hyperspectral remote sensing has already unquestionably demonstrated its strong potential for use both by the scientific community and for land management applications in coastal areas.

6.6. References

Almar, R., Bergsma, E.W., Maisongrande, P., de Almeida, L.P.M. (2019). Wave-derived coastal bathymetry from satellite video imagery: A showcase with pleiades persistent mode. *Remote Sensing of Environment*, 231, 111263.

Anflo, K. and Möllerberg, R. (2009). Flight demonstration of new thruster and green propellant technology on the prisma satellite. *Acta Astronaut*, 65, 1238–1249.

Bajjouk, T., Mouquet, P., Ropert, M., Quod, J.-P., Hoarau, L., Bigot, L., Le Dantec, N., Delacourt, C., Populus, J. (2019). Detection of changes in shallow coral reefs status: Towards a spatial approach using hyperspectral and multispectral data. *Ecological Indicators*, 96, 174–191.

Belluco, E., Camuffo, M., Ferrari, S., Modenese, L., Silvestri, S., Marani, A., Marani, M. (2006). Mapping salt-marsh vegetation by multispectral and hyperspectral remote sensing. *Remote Sensing of Environment*, 105(1), 54–67.

Berk, A., Bernstein, L., Anderson, G., Acharya, P., Robertson, D., Chetwynd, J., Adler-Golden, S. (1998). Modtran cloud and multiple scattering upgrades with application to aviris. *Remote Sensing of Environment*, 65(3), 367–375.

Bertels, L., Vanderstraete, T., Van Coillie, S., Knaeps, E., Sterckx, S., Goossens, R., Deronde, B. (2008). Mapping of coral reefs using hyperspectral CASI data; a case study: Fordata, Tanimbar, Indonesia. *International Journal of Remote Sensing*, 29(8), 2359–2391.

Bioucas-Dias, J.M., Plaza, A., Dobigeon, N., Parente, M., Du, Q., Gader, P., Chanussot, J. (2012). Hyperspectral unmixing overview: Geometrical, statistical, and sparse regression-based approaches. *IEEE Journal of Selected Topics in Applied Earth Observations and Remote Sensing*, 5(2), 354–379.

Blaaberg, S., Løke, T., Baarstad, I., Fridman, A., Koirala, P. (2014). HySpex ODIN-1024: A new high-resolution airborne HSI system. *Infrared Technology and Applications XL. International Society for Optics and Photonics*, Baltimore.

Brando, V., Anstee, J., Wettle, M., Dekker, A., Phinn, S., Roelfsema, C. (2009). A physics based retrieval and quality assessment of bathymetry from suboptimal hyperspectral data. *Remote Sens. Environ.*, 113, 755–770.

Briottet, X., Marion, R., Carrere, V., Jacquemoud, S., Chevrel, S., Prastault, P., D'oria, M., Gilouppe, P., Hosford, S., Lubac, B. (2011). Hypxim: A new hyperspectral sensor combining science/defence applications. In *IEEE 3rd Workshop on Hyperspectral Image and Signal Processing: Evolution in Remote Sensing (WHISPERS)*, Lisbon, 1–4.

Briottet, X., Adeline, K., Bajjouk, T., Carrère, V., Chami, M., Constans, Y., Derimian, Y., Dupiau, A., Dumont, M., Doz, S. et al. (2024). End-to-end simulations to optimize imaging spectroscopy mission requirements for seven scientific application domains, *ISPRS Open Journal of Photogrammetry and Remote Sensing*, 12, 100060.

Brown, W.L., Polcyn, F.C., Stewart, S.R. (1971). A method for calculating water depth attenuation coefficients and bottom reflectance characteristics (water depth attenuation coefficients and bottom reflectance characteristics from large area multispectral scanner measurements for discharge and concentrations monitoring). In *7th International Symposium on Remote Sensing of Environment*, University of Michigan, Ann Arbor, 663–682.

Caballero, I. and Stumpf, R. (2019). Retrieval of nearshore bathymetry from Sentinel-2a and 2b satellites in South Florida coastal waters. *Estuar. Coast. Shelf Sci.*, 226, 106277.

Campbell, G., Phinn, S., Dekker, A., Brando, V. (2011). Remote sensing of water quality in an Australian tropical freshwater impoundment using matrix inversion and MERIS images. *Remote Sens. Environ.*, 115, 2402–2414.

Casal, G., Sánchez-Carnero, N., Sánchez-Rodríguez, E., Freire, J. (2011). Remote sensing with SPOT-4 for mapping kelp forests in turbid waters on the south European Atlantic shelf. *Estuarine, Coastal and Shelf Science*, 91(3), 371–378.

Chami, M., Lafrance, B., Fougnie, B., Chowdhary, J., Harmel, T., Waquet, F. (2015). Osoaa: A vector radiative transfer model of coupled atmosphere-ocean system for a rough sea surface application to the estimates of the directional variations of the water leaving reflectance to better process multi-angular satellite sensors data over the ocean. *Optics Express*, 23(21), 27829–27852.

Chang, G., Mahoney, K., Briggs-Whitmire, A., Kohler, D., Mobley, C., Lewis, M., Moline, M., Boss, E., Kim, M., Philpot, W. et al. (2004). The new age of hyperspectral oceanography. *Oceanography*, 17(2), 16, 17–23.

Clark, R., Fay, T., Walker, C. (1987). Bathymetry calculations with Landsat 4 TM imagery under a generalized ratio assumption. *Appl. Opt.*, 26, 4036–4038.

Convention of the United Nations (1992). Convention on biological diversity, key international instrument for sustainable development.

Cox, C. and Munk, W. (1954). Measurement of the roughness of the sea surface from photographs of the sun's glitter. *J. Opt. Soc. Am.*, 44, 838–850.

Dekker, A., Phinn, S., Anstee, J., Bissett, P., Brando, V., Casey, B., Fearns, P., Hedley, J., Klonowski, W., Lee, Z. (2011). Intercomparison of shallow water bathymetry, hydro-optics, and benthos mapping techniques in Australian and Caribbean coastal environments. *Limnol Ocean Methods*, 9, 396–425.

Fearns, P., Klonowski, W., Babcock, R., England, P., Phillips, J. (2011). Shallow water substrate mapping using hyperspectral remote sensing. *Continental Shelf Research*, 31(12), 1249–1259.

Filippi, A. and Kubota, T. (2008). Introduction of spatial smoothness constraints via linear diffusion for optimization-based hyperspectral coastal ocean remote-sensing inversion. *J. Geophys. Res. Oceans*, 113(C3).

Fornes, A., Basterretxea, G., Orfila, A., Jordi, A., Álvarez, A., Tintoré, J. (2006). Mapping *Posidonia oceanica* from IKONOS. *ISPRS Journal of Photogrammetry and Remote Sensing*, 60(5), 315–322.

Gaddis, L., Soderblom, L., Kieffer, H., Becker, K., Torson, J., Mullins, K. (1996). Decomposition of AVIRIS spectra: Extraction of surface-reflectance, atmospheric, and instrumental components. *IEEE Trans. Geosci. Remote Sens.*, 34, 163–178.

Gao, B., Davis, C., Goetz, A. (2006). A review of atmospheric correction techniques for hyperspectral remote sensing of land surfaces and ocean color. In *IEEE International Symposium on Geoscience and Remote Sensing*, Denver, CO.

Gege, P. (2004). The water color simulator WASI: An integrating software tool for analysis and simulation of optical in situ spectra. *Comput. Geosci.*, 30, 523–532.

Gergely, M. and Zibordi, G. (2014). Assessment of AERONET-OC LWN uncertainties. *Metrologia*, 51, 40–47.

Giardino, C., Candiani, G., Bresciani, M., Lee, Z., Gagliano, S., Pepe, M. (2012). Bomber: A tool for estimating water quality and bottom properties from remote sensing images. *Comput. Geosci.*, 45, 313–318.

Giardino, C., Brando, V., Gege, P., Pinnel, N., Hochberg, E., Knaeps, E., Reusen, I., Doerffer, R., Bresciani, M., Braga, F. (2019). Imaging spectrometry of inland and coastal waters: State of the art, achievements and perspectives. *Surv. Geophys.*, 40, 401–429.

Gohin, F., Druon, J., Lampert, L. (2002). A five channel chlorophyll concentration algorithm applied to SeaWiFS data processed by SeaDAS in coastal waters. *Int. J. Remote Sens.*, 23, 1639–1661.

Goodman, J. and Ustin, S. (2007). Classification of benthic composition in a coral reef environment using spectral unmixing. *J. Appl. Remote. Sens.*, 1(1), 011501.

Goodman, J., Lee, Z., Ustin, S. (2008). Influence of atmospheric and sea-surface corrections on retrieval of bottom depth and reflectance using a semi-analytical model: A case study in Kaneohe Bay, Hawaii. *Appl. Opt.*, 47, F1–F11.

Guanter, L., Kaufmann, H., Segl, K., Foerster, S., Rogass, C., Chabrillat, S., Kuester, T., Hollstein, A., Rossner, G., Chlebek, C. (2015). The EnMAP spaceborne imaging spectroscopy mission for earth observation. *Remote Sens.*, 7, 8830–8857.

Harmel, T., Chami, M., Tormos, T., Reynaud, N., Danis, P.-A. (2018). Sunglint correction of the multi-spectral instrument (MSI)-SENTINEL-2 imagery over inland and sea waters from SWIR bands. *Remote Sens. Environ.*, 204, 308–321.

Hedley, J., Harborne, A., Mumby, P. (2005). Simple and robust removal of sun glint for mapping shallow-water benthos. *Int. J. Remote Sens.*, 26, 2107–2112.

Hedley, J., Roelfsema, C., Phinn, S. (2009). Efficient radiative transfer model inversion for remote sensing applications. *Remote Sens. Environ.*, 113, 2527–2532.

Hedley, J., Russell, B.J., Randolph, K., Pérez-Castro, M., Vásquez-Elizondo, R.M., Enríquez, S., Dierssen, H.M. (2017). Remote sensing of seagrass leaf area index and species: The capability of a model inversion method assessed by sensitivity analysis and hyperspectral data of Florida Bay. *Frontiers in Marine Science*, 4, 362. doi: 10.3389/fmars.2017.00362.

Hick, P. and Scoones, R. (1988). Satellite data use for studies of benthic and bathymetric characteristics of sub-sea stratigraphy in the eighty mile beach region of western Australia – Preliminary report for the Western Australian Fisheries Department of a collaborative research project between CSIRO and Broome Pearls. Technical Report.

Hochberg, E., Atkinson, M., Andréfouët, S. (2003). Spectral reflectance of coral reef bottom-types worldwide and implications for coral reef remote sensing. *Remote Sens. Environ.*, 85, 159–173.

Holden, H. and LeDrew, E. (2001). Hyperspectral discrimination of healthy versus stressed corals using in situ reflectance. *J. Coast. Res.*, 17, 850–858.

Huang, R., Yu, K., Wang, Y., Wang, W., Mu, L., Wang, J. (2018). Method to design a live coral cover sensitive index for multispectral satellite images. *Opt. Express*, 26, A374–A397.

Jagalingam, P., Akshaya, B., Hegde, A. (2015). Bathymetry mapping using landsat 8 satellite imagery. *Procedia Eng.*, 116, 560–566.

Janesick, J.R. (2007). *Photon Transfer*. SPIE Press, Bellingham, WA.

Jay, S. and Guillaume, M. (2011). Estimation of water column parameters with a maximum likelihood approach. In *IEEE 3rd Workshop on Hyperspectral Image and Signal Processing: Evolution in Remote Sensing (WHISPERS)*, Lisbon, 1–4.

Jordan, A., Lawler, M., Halley, V., Barrett, N. (2005). Seabed habitat mapping in the Kent Group of islands and its role in marine protected area planning. *Aquat. Conserv. Mar. Freshw. Ecosyst.*, 15, 51–70.

Kaufmann, H., Segl, K., Chabrillat, S., Hofer, S., Stuffler, T., Mueller, A., Richter, R., Schreier, G., Haydn, R., Bach, H. (2006). EnMAP a hyperspectral sensor for environmental mapping and analysis. In *IEEE International Conference on Geoscience and Remote Sensing Symposium*, Denver, CO.

Kay, S., Hedley, J., Lavender, S. (2009). Sun glint correction of high and low spatial resolution images of aquatic scenes: A review of methods for visible and near-infrared wavelengths. *Remote Sens.*, 1, 697–730.

Keith, D., Schaeffer, B., Lunetta, R., Gould Jr, R., Rocha, K., Cobb, D. (2014). Remote sensing of selected water-quality indicators with the hyperspectral imager for the coastal ocean (HICO) sensor. *Int. J. Remote Sens.*, 35, 2927–2962.

Klonowski, W., Fearns, P., Lynch, M. (2007). Retrieving key benthic cover types and bathymetry from hyperspectral imagery. *J. Appl. Remote Sens.*, 1, 011505.

Kutser, T., Pierson, D., Kallio, K., Reinart, A., Sobek, S. (2005). Mapping lake cdom by satellite remote sensing. *Remote Sens. Environ.*, 94, 535–540.

Kutser, T., Miller, I., Jupp, D.L.B. (2006). Mapping coral reef benthic substrates using hyperspectral space-borne images and spectral libraries. *Estuarine Coastal Shelf Sci.*, 70, 449–460.

Kutser, T., Vahtmäe, E., Praks, J. (2009). A sun glint correction method for hyperspectral imagery containing areas with non-negligible water leaving NIR signal. *Remote Sens. Environ.*, 113, 2267–2274.

Lafon, V., Froidefond, J., Lahet, F., Castaing, P. (2002). Spot shallow water bathymetry of a moderately turbid tidal inlet based on field measurements. *Remote Sens. Environ.*, 81, 136–148.

Lee, Z., Carder, K., Mobley, C., Steward, R., Patch, J. (1998). Hyperspectral remote sensing for shallow waters. I. A semianalytical model. *Appl. Opt.*, 37, 6329–6338.

Lee, Z., Carder, K., Mobley, C., Steward, R., Patch, J. (1999). Hyperspectral remote sensing for shallow waters. 2. Deriving bottom depths and water properties by optimization. *Appl. Opt.*, 38, 3831–3843.

Lorthiois, T., Doxaran, D., Chami, M. (2012). Daily and seasonal dynamics of suspended particles in the Rhône River plume based on remote sensing and field optical measurements. *Geo-Marine Letters*, 32(2), 89–101.

Lyzenga, D. (1981). Remote sensing of bottom reflectance and water attenuation parameters in shallow water using aircraft and landsat data. *Int. J. Remote Sens.*, 2, 71–82.

Lyzenga, D., Malinas, N., Tanis, F. (2006). Multispectral bathymetry using a simple physically based algorithm. *IEEE Trans. Geosci. Remote Sens.*, 44, 2251–2259.

Martin, J., Eugenio, F., Marcello, J., Medina, A. (2016). Automatic sun glint removal of multispectral high-resolution worldview-2 imagery for retrieving coastal shallow water parameters. *Remote Sens.*, 8(1), 37.

Matthew, M., Adler-Golden, S., Berk, A., Felde, G., Anderson, G., Gorodetzky, D., Paswaters, S., Shippert, M. (2002). Atmospheric correction of spectral imagery: Evaluation of the FLAASH algorithm with AVIRIS data. In *IEEE Applied Imagery Pattern Recognition Workshop Proceedings*, Washington, DC.

Mélin, F. (2019). *Uncertainties in Ocean Colour Remote Sensing. Reports of the International Ocean Colour Coordinating Group*. International Ocean Colour Coordinating Group (IOCCG), Dartmouth, NS. doi: 10.25607/OBP-696.

Minghelli-Roman, A., Goreac, A., Mathieu, S., Spigai, M., Gouton, P. (2009). Comparison of bathymetric estimation using different satellite images in coastal sea waters. *Int. J. Remote Sens.*, 30, 5737–5750.

Mishra, D., Narumalani, S., Rundquist, D., Lawson, M., Perk, R. (2007). Enhancing the detection and classification of coral reef and associated benthic habitats: A hyperspectral remote sensing approach. *Journal of Geophysical Research: Oceans*, 112, C8.

Mobley, C., Boss, E., Roesler, C. (2011). *Ocean Optics Web Book* [Online]. Available at: https://www.oceanopticsbook.info/view/radiative-transfer-theory/level-2/hydrolight.

Morel, A. and Gordon, H. (1980). Report of the working group on water color. *Boundary-Layer Meteorology*, 18, 343–355.

Moses, W., Sterckx, S., Montes, M., Keukelaere, D.L., Knaeps, E. (eds) (2017). Atmospheric correction for inland waters. In *Bio-optical Modeling and Remote Sensing of Inland Waters*. Elsevier, Amsterdam.

Mumby, P., Green, E., Edwards, A., Clark, C. (1999). The cost-effectiveness of remote sensing for tropical coastal resources assessment and management. *J. Environ. Manag.*, 55, 157–166.

Nechad, B., Ruddick, K., Park, Y. (2010). Calibration and validation of a generic multisensor algorithm for mapping of total suspended matter in turbid waters. *Remote Sens. Environ.*, 114, 854–866.

Nieto, P. (2013). Classifying benthic habitats and deriving bathymetry at the Caribbean Netherlands using multispectral imagery. Case study of St. Eustatius. Master's Thesis, Wageningen University & Research, Wageningen.

O'Reilly, J.E., Maritorena, S., Siegel, D.A., O'Brien, M.C., Toole, D., Mitchell, B.G., Kahru, M., Chavez, F.P., Strutton, P., Cota, G.F. et al. (2000). Ocean color chlorophyll a algorithms for SeaWiFS. Technical Report, Marine Science Faculty Publications, College of Marine Science, University of South Florida.

Petit, T., Bajjouk, T., Mouquet, P., Rochette, S., Vozel, B., Delacourt, C. (2017). Hyperspectral remote sensing of coral reefs by semi-analytical model inversion – Comparison of different inversion setups. *Remote Sens. Environ.*, 190, 348–365.

Philpot, W. (1989). Bathymetric mapping with passive multispectral imagery. *Appl. Opt.*, 28, 1569–1578.

Philpot, W. (2007). Estimating atmospheric transmission and surface reflectance from a glint-contaminated spectral image. *IEEE Trans. Geosci. Remote Sens.*, 45, 448–457.

Piazzi, L., Acunto, S., Cinelli, F. (2000). Mapping of *Posidonia oceanica* beds around Elba Island (western Mediterranean) with integration of direct and indirect methods. *Oceanol. Acta*, 23, 339–346.

Pluquet, F. and Ehrhold, A. (2009). Une nouvelle stratégie d'étude des habitats marins littoraux au moyen de la vedette acoustique v/o haliotis. Technical Report, Ifremer.

Populus, J. (2002). Altimétrie par lidar aéroporté et modèles numériques de terrain. Technical Report, Ifremer.

Richter, R. and Schläpfer, D. (2012). Atmospheric/topographic correction for airborne imagery. ATCOR-4 user guide 565–02. Technical Report.

Sauriau, P., Pichocki-Seyfried, C., Walker, P., De Montaudouin, X., Palud, C., Heral, M. (1998). Crepidula fornicata L. (mollusque, gastéropode) en baie de marennes-oléron : cartographie des fonds par sonar à balayage latéral et estimation du stock. *Oceanol. Acta*, 21, 353–362.

Sturm, B. (1981). Ocean colour remote sensing and quantitative retrieval of surface chlorophyll in coastal waters using Nimbus CZCS data. In *Oceanography from Space*, Gower, J.F.R. (ed.). Springer, New York.

Vahtmäe, E., Paavel, B., Kutser, T. (2020). How much benthic information can be retrieved with hyperspectral sensor from the optically complex coastal waters? *Journal of Applied Remote Sensing*, 14(1), 016504–016504. doi: 10.1117/1.JRS.14.016504.

Vermote, E., Tanre, D., Deuze, J., Herman, M., Morcrette, J. (1997). Second simulation of the satellite signal in the solar spectrum, 6s: An overview. *IEEE Trans. Geosci. Remote Sens.*, 35, 675–686.

Wang, M. and Bailey, S. (2001). Correction of sun glint contamination on the SeaWiFS ocean and atmosphere products. *Appl. Opt.*, 40, 4790–4798.

Zhang, H. and Wang, M. (2010). Evaluation of sun glint models using MODIS measurements. *J. Quant. Spectrosc. Radiat. Transf*, 111, 492–506.

Zibordi, G., Mélin, F., Berthon, J.F., Holben, B., Slutsker, I., Giles, D., Fabbri, B.E. (2009). AERONET-OC: A network for the validation of ocean color primary products. *Journal of Atmospheric and Oceanic Technology*, 26(8), 1634–1651.

Zibordi, G., Mélin, F., Berthon, J.-F., Canuti, E. (2013). Assessment of MERIS ocean color data products for European seas. *Ocean Science Discussions*, 9(3), 521–533.

7

Deep-learning Analysis of Cherenkov Telescope Array Images

Mikaël JACQUEMONT[1,2], Thomas VUILLAUME[3], Alexandre BENOIT[2], Gilles MAURIN[3] and Patrick LAMBERT[2]

[1] *LAPP, Université Grenoble Alpes, Annecy, France*

[2] *LISTIC, Université Savoie Mont Blanc, Annecy, France*

[3] *LAPP, Université Grenoble Alpes, Université Savoie Mont Blanc, Annecy, France*

7.1. Gamma astronomy

Astronomy is a scientific field that creates very large amounts of data from which researchers try to characterize or model complex phenomena. As such, astronomy requires efficient methodologies for data processing and assimilation.

7.1.1. *Gamma radiation and observation method*

Gamma-ray astronomy consists of the observation of the most energetic photons produced by violent astrophysical phenomena. These are mainly emitted during the interaction of accelerated particles, called cosmic rays, with ambient matter or

For a color version of all the figures in this chapter, see www.iste.co.uk/yan/remotesensing.zip.

Inversion and Data Assimilation in Remote Sensing,
coordinated by Yajing YAN.

electromagnetic fields. Their study allows, for example, to better understand the laws that underlie the creation of stars and the evolution of galaxies. It is also employed to explore new physics such as dark matter or the violation of Lorentz invariance. The detection of this radiation from Earth is indirectly achieved, by observing Cherenkov light emitted by the electromagnetic shower that they generate when penetrating the atmosphere, as illustrated in Figure 7.1.

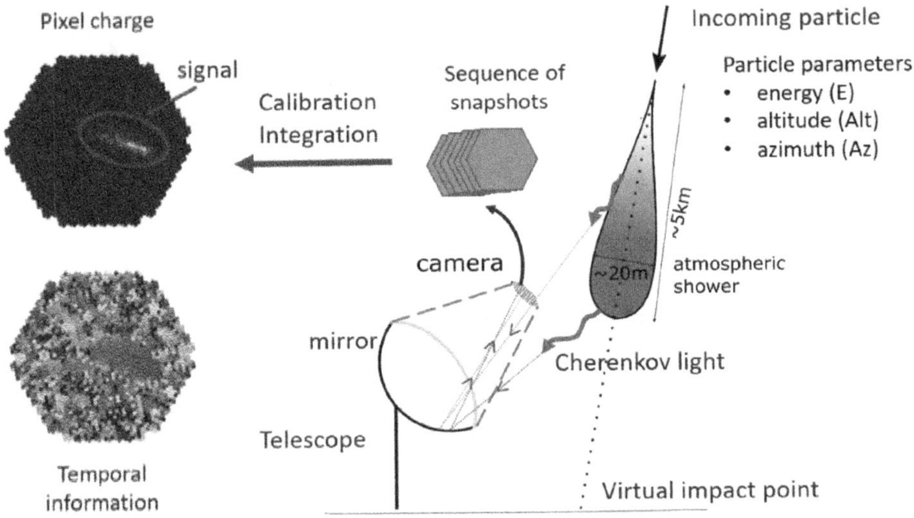

Figure 7.1. *Principle of the Cherenkov effect telescope, or Imaging Atmospheric Cherenkov Telescope (IACT)*

These observations are made possible through the use of specific telescopes such as the *Imaging Atmospheric Cherenkov Telescopes* (IACT). Cherenkov light thus collected by ultra-sensitive cameras leaves a characteristic trace, whose development only lasts a few nanoseconds. The analysis of gamma radiation then consists of its *separation from background noise* – cosmic rays, mainly dominated by protons (classification task of the types of particles) – and *reconstruction of its energy and incident direction* (regression tasks).

This analysis is complex, because cosmic rays can produce very similar images to those of gammas (see the examples in Figure 7.2) and the signal-to-noise ratio is typically less than 1/1,000. In addition, the images are largely composed of noise originating from both the sensor electronics and the bright background of the night sky. Finally, atmospheric gamma ray detection, especially in the context of a single telescope, introduces degeneracies in the parameters to be reconstructed. There indeed exists a strong interdependence between, on the one hand, the energy, direction and virtual point of impact on the ground of the incident particle and on the other hand, the images produced in the telescope camera. For a given direction and point of impact, the signal strength is strongly related to the gamma ray energy (Völk and Bernlöhr 2009). The position and shape of the signal produced in the camera depends on the direction and impact point. Gamma radiation analysis is therefore multitasking in nature.

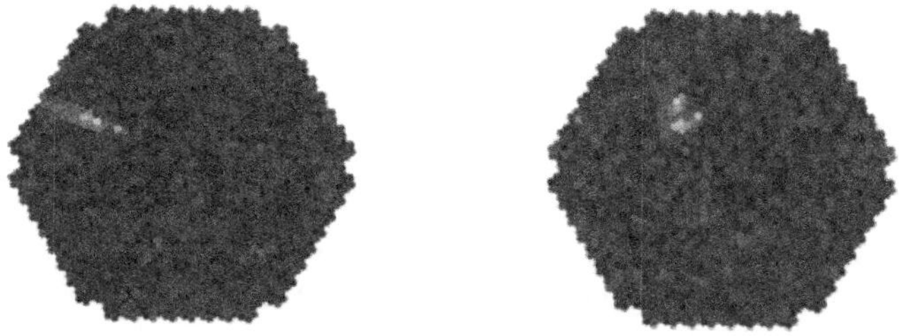

a) b)

Figure 7.2. *(a) 0.5 TeV energy gamma event; (b) 27 GeV energy proton event*

Ground-based gamma-ray astronomy is relatively young, and since Whipple's first detection of the Crab Nebula in 1989 (Weekes et al. 1989), further observatories have been built. They are mainly designed as telescope arrays to take advantage from stereoscopy. These include the *Cherenkov telescope array* (CTA)[1] that embodies the new generation of observatories of the universe at very high energy from 20 GeV to 300 TeV. Comprising more than 100 telescopes of three different sizes distributed over two sites, the CTA will have a sensitivity 10 times better than current experiments, such as the *High Energy Stereoscopic System* (H.E.S.S.)[2], the Magic telescopes[3] or the *Very Energetic Radiation Imaging Telescope Array System* (Veritas)[4], while improving reconstruction accuracy. In addition, once completed, the CTA will produce a huge amount of data, more than 200 PB/year, which will have to be analyzed in real time before being reduced and compressed for transfer and storage. The data acquired will thus be stored and processed again each year in order to benefit from improvements in analysis methods.

Although the CTA is being constructed, a prototype of the larger telescopes, known as *large-sized telescopes* (LST), called LST1, was inaugurated in the fall of 2018

1 Available at: https://www.cta-observatory.org/.

2 Available at: http://www.mpi-hd.mpg.de/hfm/HESS/pages/about/telescopes.

3 Available at: http://www.magic.iac.es/.

4 Available at: https://veritas.sao.arizona.edu/.

and began providing usable data in spring 2020. LSTs have been designed to have optimized sensitivity in the lower part of the spectrum (up to a few TeV) of very high energy gamma rays. This configuration makes them suitable for the study of transient and extragalactic phenomena. LST1 is also particularly important for the CTA collaboration, as it will allow astrophysicists to test their analytical tools. Ground truth is indeed impossible to obtain for the real data produced by telescopes. As a result, analysis models are developed using a large amount of high-quality simulation data.

7.1.2. *Analysis methods for Cherenkov images*

In the case of LST1, both actual and simulated raw data consist of spatiotemporal cubes of 40 samples of 1 ns, each called waveforms. In both cases, these *waveforms* are calibrated and integrated using the lstchain v0.6.3 library (López-Coto et al. 2021) in its standard configuration. The input data of the analysis models are then composed of two channels, the first representing the amount of photoelectrons received per pixel (the intensity) and the second containing temporal information on the development of the shower, also defined per pixel. More precisely, this temporal information represents, for each pixel, the temporal position of the peak intensity during the development of the shower.

Currently, the most widely used analysis method is based on the extraction of the geometric parameters of the shower proposed by A.M. Hillas (Hillas 1985) and the analysis of these parameters by a multivariate method (Bock et al. 2004; Albert et al. 2008; Ohm et al. 2009; Fiasson et al. 2010). This method is called *Hillas + RF* later in this chapter. As shown in Figure 7.3, the Hillas method assumes that the signal produced by a gamma ray in the telescope camera is an ellipsoid. After a background noise suppression step, the moments of the ellipse are characterized. In addition to the total intensity of the shower, the position of the ellipse centroid (its distance from the center of the camera as well as the azimuth angle φ), the major and minor half-axes and the angle α formed by the major axis of the ellipse and the axis connecting its centroid to the center of the camera are determined. These parameters are then fed into a *Random Forest* (RF)-based *machine learning* algorithm to perform the reconstruction of the event. For a stereoscopic observation of the events (by an array of telescopes), the incident direction of the detected gamma ray can be geometrically determined: the images produced by each telescope are superimposed in an altitude-azimuth frame of reference; the source position then corresponds to the intersection of the major axes of the ellipsoids.

In reality, the most efficient method, called *template analysis*, consists of a pixel-to-pixel comparison, using a likelihood function, of the images produced by the telescopes with a very large database of models (de Naurois and Rolland 2009;

Parsons and Hinton 2014), the best comparison then providing the desired characteristics. However, there are limitations to the application of these methods to the CTA, either in terms of sensitivity (Hillas + RF) or computational capabilities (*template analysis*); the number of comparisons to be performed can become extremely significant. It is therefore necessary to explore other avenues.

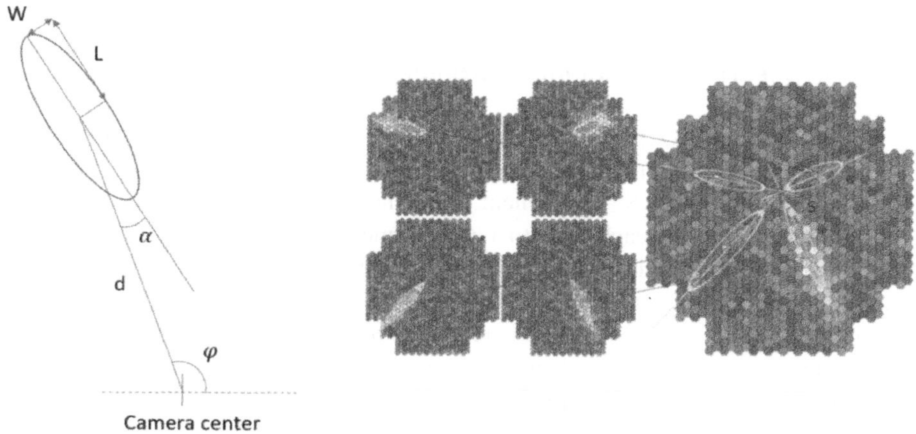

a) b)

Figure 7.3. *(a) Moment extraction from the ellipse produced by a gamma ray. (b) Stereoscopy leads to geometrically determining the direction of the incident particle (source: Analysis Methods for Gamma-ray Astronomy – Scientific Figure on ResearchGate.)*

Relying on recent advances in artificial neural networks, we propose in this chapter a new deep learning approach to analyzing CTA data, especially those from LST1. We present a multitasking architecture, γ-PhysNet, inspired by physics that performs the complete reconstruction of gamma events. It benefits from an attention mechanism, making it more robust to initialization conditions. From the simulation data employed to prepare the analysis algorithms, γ-PhysNet achieves significantly better results, both in sensitivity and spatial resolution, than the standard Hillas + RF method. Low-energy sensitivity gain could help improve the study of transient phenomena. Moreover, we present a preliminary analysis of the proposed network using a visual explainability method to explain its behavior in detail.

7.2. Deep neural networks

Since the triumph of AlexNet (Krizhevsky et al. 2012) at the ILSVRC2012 challenge (Russakovsky et al. 2015), deep learning has emerged as the dominant approach to solving many computer-based vision problems such as image classification (Touvron et al. 2019), semantic segmentation (Yuan et al. 2020) or object detection (Zhang et al. 2020). The field of astrophysics is no exception (Kim and Brunner 2016; Brunel et al. 2019). More specifically, deep learning has been explored for the analysis of images produced by IACTs from muon analysis (Feng et al. 2016) to gamma event reconstruction.

7.2.1. *Deep learning for gamma astronomy*

Holch et al. (2017) proposed a shallow convolutional neural network (three convolution layers) for the analysis of stereoscopic data acquired by the four telescopes of the H.E.S.S. experiment. The methodology adopted consisted of optimizing a model for each task to be performed: gamma/proton classification, energy regression and incident particle direction regression. To combine information from the four telescopes, the images produced were summed in a pixelwise manner to form only one image, analogous to the direction reconstruction in the Hillas + RF method. Despite the fact that their model performed well for separating gamma events from background noise, Holch et al. (2017) provided only preliminary results for regression tasks. Taking this work a step further, Shilon et al. (2019) proposed the CRNN model for gamma/proton classification, integrating a recurring LSTM-type layer (Cheng et al. 2016). Placed after three convolution layers similar to the Holch model, the purpose of this recurring layer was to improve stereoscopy processing. With this strategy, instead of being summed, the images from the four telescopes were presented to the array as a sequence, ordered according to the total amplitude of the images. In contrast, for direction reconstruction, the authors followed a different approach. The images from the four telescopes were combined into a multi-channel representation, one image per channel. The proposed model was also deeper, consisting of five sets of two convolutions, followed by subsampling. However, the good performance obtained on simulation data was not replicated in real data analysis. Moreover, this work did not propose energy reconstruction. In order to solve the performance difference observed between simulation and real data analyses, Parsons and Ohm (2020) proposed to combine the latent representation produced by the convolutional part of the model with the geometric parameters extracted by the standard method. However, the improvement brought by their method compared to the studies by Shilon et al. (2019) was not discussed.

Preliminary work has been carried out focusing on the CTA. Nieto et al. (2017) investigated the separation of gamma events from protons for a single telescope relying on two very deep arrays, based on recent architectures initially proposed for multimedia image recognition problems, namely a ResNet-50 (He et al. 2016b) and an Inception V3 (Szegedy et al. 2015). These two models achieved similar results, however much worse than those obtained with a standard analysis chain such as Hillas + RF. On the other hand, Mangano et al. (2018) tested much simpler convolutional networks for particle classification and energy and direction regression, with each task performed using a dedicated model. Utilizing the combination of images method proposed by Holch et al. (2017) and carried out in a very constrained context of high data selection, this stereoscopic data analysis did not provide better results than the standard method.

Although all these works presented promising results, especially for gamma/proton classification, all addressed the different tasks of gamma event reconstruction as independent problems. Their single task approach did not take into account the strong interdependence of the parameters to be reconstructed.

7.2.2. *Multitasking learning*

The interdependence of the parameters to be reconstructed makes the reconstruction of gamma events from images originating from IACT multitasking by nature. Multitask learning (Caruana 1997) is a paradigm of learning all tasks simultaneously using a common model, rather than dedicating a different model to each task. The multitasking approach increases the generalization capacity of the model by helping it focus on characteristics that are relevant to all tasks. This approach avoids overspecialization for a particular task of the shared parameters (Baxter 1997).

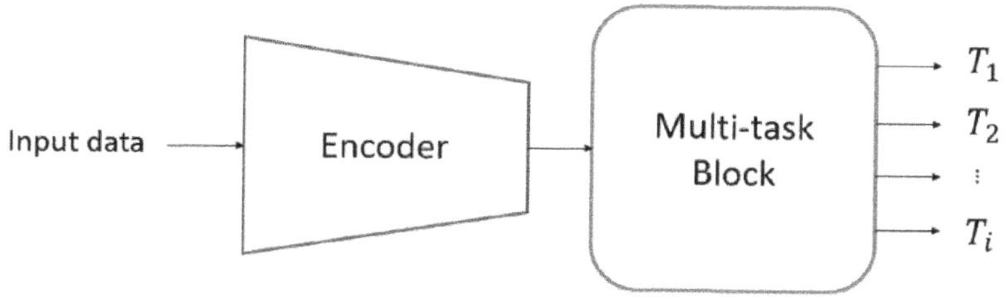

Figure 7.4. *Example of hard parameter sharing architecture. The encoder is shared by all tasks. The multitasking block defines the specific path for each task*

The most commonly used network topology, named *hard parameter sharing*, consists, as shown in Figure 7.4, of sharing between all the tasks part of the model, usually the encoder (Kendall et al. 2018; Luvizon et al. 2018; Ren and Jae Lee 2018) or its first layers (Iizuka et al. 2016). The task-specific parts of the network are usually composed of fully connected layers, but can also be convolution layers for certain computer vision problems such as semantic segmentation (Xu et al. 2018). The other possible model topology is called *soft parameter sharing*. In this case, each task is learned by a different network. The addition of additional layers then makes it possible to share information between different (Cao et al. 2018) tasks or to constrain the weights of certain specific layers of each network to similar distributions, for example, by regularizing the norm of parameter tensors concatenated per layer between the different networks (Yang and Hospedales 2017). An example of a *soft parameter sharing* architecture is given in Figure 7.5.

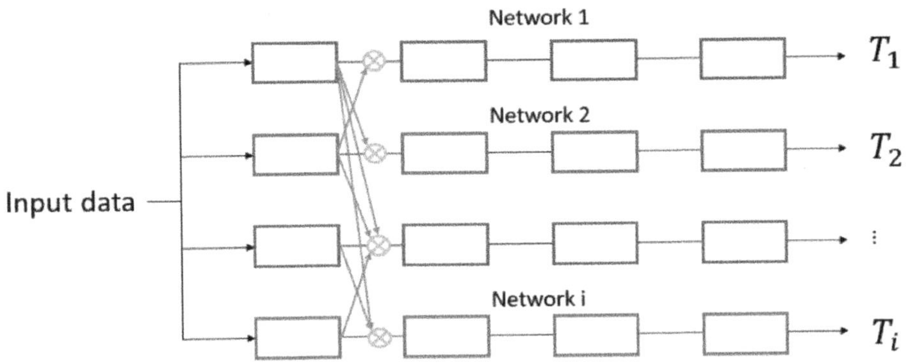

Figure 7.5. *Example of soft parameter sharing architecture*

COMMENT ON FIGURE 7.5.– *Each task is learned by a different network, but all networks have the same structure. Information is exchanged between tasks by means of a specific operator that combines the maps of characteristics produced by the first stage of each network. For more clarity, not all flows from these feature maps are shown.*

Although, in theory, multitasking reduces the risk of overfitting, it is in practice fundamental that the different tasks be balanced when computing the cost function in order to prevent one task that is easier to learn from taking advantage of the others. In most research on multitasking deep learning (Han et al. 2017; Kokkinos 2017; Luvizon et al. 2018; Ren and Jae Lee 2018), when specified, this balancing is manually carried out. However, this requires a significant optimization process. Nonetheless, there are adaptive methods to automatically balance task priority. Kendall et al. (2018) modeled the homoscedastic uncertainty of each task as an additional task and used it as a proxy to balance them. Following a different approach, Chen et al. (2018) proposed to weight each task so that the gradient of their individual cost function with respect to the last common layer has a similar amplitude to that of the other tasks. This method amounts to penalizing the predominant tasks to help the most difficult to learn. Guo et al. (2018) used learning progress signals, also known as key performance indicators, such as average accuracy, rather than the error computed by the cost function of each task, in order to prioritize difficult cases. This prioritization was conducted at two levels: from the point of view of the task to be learned, on the one hand, and the learning example itself, on the other hand. Sener and Koltun (2018) considered multitasking deep learning as a multiobjective optimization problem. Their method achieved Pareto optimality for weights balancing each task.

7.2.3. *Attention mechanisms*

Due to the degeneracy introduced into the parameters by the atmospheric detection of gamma rays, key information for a good event reconstruction may be found in the contour and distribution of the intensity of the signal present in the image. Attention mechanisms can help the model capture these subtle details and thus improve performance. From a defined context, these mechanisms help the model to focus on the relevant characteristics of the data. From a biological point of view, we can draw parallels with the human visual system that focuses on the relevant parts of its field of vision in order to facilitate perception (Saenz et al. 2002; Boynton 2005). In deep learning, these attention mechanisms learn the short- and long-range dependencies, depending on the context taken into account, present in the data.

Attention mechanisms have emerged in the field of natural language processing (Bahdanau et al. 2015). These are the main components of the transformer models (Vaswani et al. 2017; Devlin et al. 2019) are the ones that obtain the best performance in machine translation and image description. In recent years, attention mechanisms have become essential components of neural networks for natural language processing and also for image processing. In this area, a distinction between three types of attention can be made:

- spatial attention, or pixel-wise;
- attention between channels;
- the combination of both.

Through restricted self-attention, Parmar et al. (2018) generalized the transformer architecture for image generation. This spatial attention mechanism learns the relevant dependencies between pixels in a local neighborhood. While attention mechanisms are typically used in addition to convolution layers for computer vision applications, Parmar et al. (2019) provided a model entirely consisting of local self-attention layers for image classification and object detection. On the other hand, Wang et al. (2018) extended this local concept to global self-attention in order to consider the entire image as the context for learning dependencies between pixels. This mechanism has been applied to video classification, image segmentation and exposure estimation.

Zhang et al. (2019) adapted global self-attention to generative adversarial networks. This method is called *self-attention* (SA) later in this chapter.

Unlike spatial attention, Hu et al. (2018) proposed a mechanism for learning dependencies between the channels of images or intermediate representations (feature maps) in the neural network. This mechanism, called *squeeze-and-excitation*, identifies the relevant channels for model decision through three operations. As shown in Figure 7.6, the (i) *squeeze* operation first produces one descriptor per input data channel. Then, the (ii) *excitation* operation performs an adaptive recalibration of these descriptors, which are then used for (iii) weighting the input data channels. It should be noted that excitation acts as a bottleneck controlled by a reduction ratio, which is an important parameter of the method.

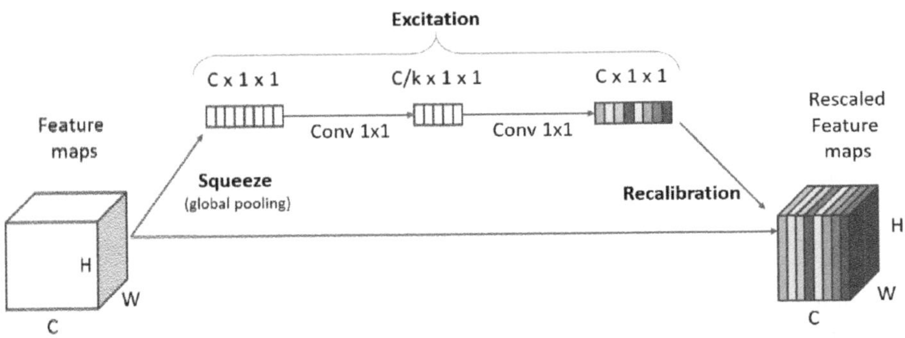

Figure 7.6. *Attention mechanism as per squeeze-and-excitation channel. The parameter k controls the bottleneck. C, H and W, respectively, represent the number of channels, the height and width of the feature maps (source: Hu et al. (2018))*

Spatial and channel attentions can be associated, as in the mechanism called *dual attention* (Sun et al. 2020). This is composed, as depicted in Figure 7.7, of an attention path per channel, consisting of a *squeeze-and-excitation* block, and a relatively simple spatial attention path. The latter involves two convolutions intended to reduce the number of channels to 1, followed by a sigmoid function generating an activation map consisting of the weight of each pixel (between 0 and 1). In order to allow the module to highlight pixels only, and not to turn them off, a scalar 1 is added to each pixel of the activation map, thus producing a spatial attention map. The latter map is then employed for recalibrating the output of the channel attention path. There are other existing mechanisms combining spatial and channel attention, such as SCA-CNN (Chen et al. 2017) designed for image caption generation, or CBAM (Woo et al. 2018), whose spatial attention path exhibits less expressiveness than that of *dual attention*.

By helping the model focus on relevant characteristics, attention mechanisms can lead to improving the robustness of neural networks. In addition, they can also provide insight into how the model makes its decisions. In particular, the observation of spatial attention maps makes it possible to identify the pixels highlighted, and therefore they play a greater role in the result provided by the model.

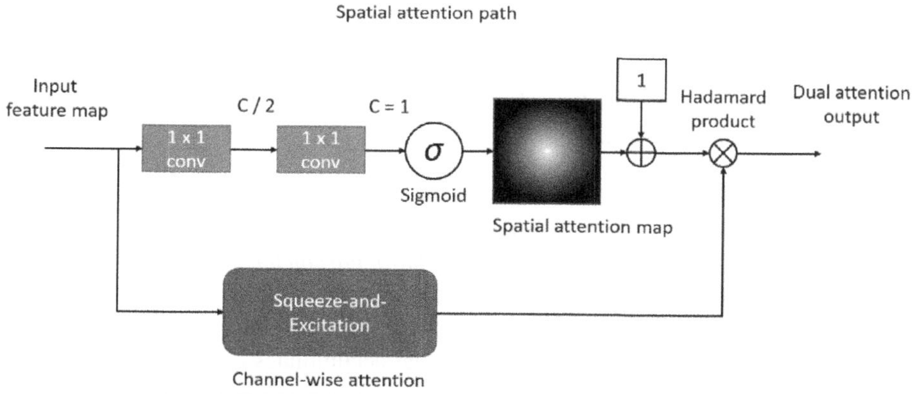

Figure 7.7. *Dual attention mechanism combining spatial attention and per channel (source: Sun et al. (2020))*

7.2.4. *Explainability of neural networks*

While traditional expert systems are highly interpretable and explainable, deep neural networks are often perceived as *black boxes*. The deep learning community's effort to make progress in these areas has considerably increased in recent years. Some methods explore the role of individual neurons or of the linear combination of neurons by ablation (Morcos et al. 2018; Zhou et al. 2018) or optimization (Olah et al. 2017). Others believe that input characteristics are important for activating a particular output. These methods produce significance maps (Cao et al. 2015; Springenberg et al. 2015; Srinivas and Fleuret 2019), also called heat maps (Selvaraju et al. 2017), for visualizing the behavior of neural networks. In the rest of this chapter, we will rely on one of them, Grad-CAM (Selvaraju et al. 2017), for the analysis of the impact of attention on the behavior of the proposed architecture. Grad-CAM is a method for localizing these features that strongly discriminates the model outputs. It produces heat maps of the most important pixels in model decision, and these maps are specific to each output of the neural network. Grad-CAM is applicable to any type of network architecture. Heat maps are computed from the last convolution layer of the model. In the context of multitasking, Grad-CAM is therefore particularly suitable for *hard parameter sharing*-type architectures, which are described in section 7.2.2 and which share a convolutional encoder between all tasks.

In addition, as already discussed in the previous section, when models involve attention mechanisms, the observation of attention maps, especially spatial ones, can also provide indications about the regions of interest on which the model focuses, and thus can help explain network decisions.

7.3. γ-PhysNet: a multitasking architecture for the complete reconstruction of gamma events

In order to achieve the full reconstruction (particle type, energy and direction) of gamma events from Cherenkov images, we propose γ-PhysNet, a deep multitasking architecture. It is an architecture based on *hard parameter sharing*, as shown in Figure 7.8. It makes it possible to reduce the number of model parameters and to reduce computation time. The speed of inference is indeed crucial for the analysis of CTA images because of the amount of data that accumulates each year and must be processed again regularly. γ-PhysNet comprises an encoder and a multitasking block inspired by the physics of the phenomenon. From the data produced by LST1, intensity maps and time information maps, γ-PhysNet performs at the same time the separation of gammas from the background noise (gamma/proton classification) and the reconstruction of the parameters of the detected particle; that is, the regression of its energy, its incident direction, but also of its virtual point of impact as an auxiliary task.

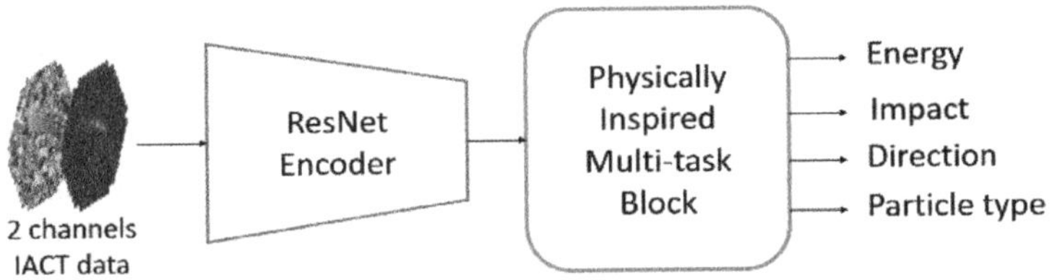

Figure 7.8. *γ-PhysNet architecture for IACT image analysis*

7.3.1. *Encoder*

An in-depth study has allowed the encoder of γ-PhysNet to be defined. We compared shallow convolutional networks (three, four and five layers), a deep neural network (the VGG-16 (Simonyan and Zisserman 2015)) as well as very deep networks, including DenseNets (Huang et al. 2017) with variable number of layers (up to 97) and different ResNets (He et al. 2016a, 2016b). For the latter, we evaluated the Cifar-10 version. The images produced by LST1 are comparable in size to that of the Cifar-10 dataset (1,855 pixels for LST1 against 1,024 for Cifar-10). ResNets globally proved they were better performing, and more specifically, the ResNet-56 Cifar-10 with full pre-activation – the normalization (BN) and activation (ReLU) layers are placed before convolution – achieved the best results in terms of training performance and stability.

We therefore adapt the convolutional part of this ResNet-56 as an encoder in γ-PhysNet. In fact, since the images generated by the LST1 have hexagonal pixels, in order to avoid an additional preprocessing step, we implemented the γ-PhysNet encoder with indexed convolutions[5] (Jacquemont et al. 2019). These can be used to apply convolution to any type of mesh while respecting the actual neighborhood of the pixels. In addition, the value of resampling images according to a Cartesian grid was not clearly demonstrated for the performance of the model (Nieto et al. 2019).

7.3.2. *Multitasking block*

The design of the multitasking block of γ-PhysNet, illustrated in Figure 7.9, is directly inspired by the physics of the tasks to be solved. This block, composed of fully connected layers, is indeed divided into two subnets, a network based on global characteristics (called *global feature network*) and another based on local characteristics (*local feature network*).

The global subnet is dedicated to energy regression. As a matter of fact, this parameter can be considered global, because for a given direction and point of impact, the total number of photoelectrons captured by the camera is approximately proportional to the energy of the gamma particle that generated the shower (Völk and Bernlöhr 2009). This subnet therefore starts with a global average pooling reducing the feature maps produced by the encoder to one descriptor per channel. This is followed by a fully connected 256-neuron layer supplemented by a ReLU-type activation and a last fully connected layer that produces energy regression.

On the other hand, gamma/proton classification and direction and point of impact regression rely on local and spatial information. As a first approximation:

– the signal position in the image depends on the direction;

– its orientation depends on the point of impact;

– its shape (elongated or not) depends on both the direction and the point of impact;

– The shape of its contour as well as its intensity distribution depend on the type of particle.

5 Available at: https://github.com/IndexedConv/IndexedConv.

To make use of these dependencies, the local subnet retains all the features of the maps produced by the encoder by vectorizing them. Then, the gamma/proton classification is carried out directly from this vector using a fully connected layer. The regressions of the direction and point of impact share a first fully connected 256-neuron layer, followed by a ReLU-based activation. Finally, each task is performed using a fully connected two-neuron layer, the direction being reconstructed by its altitude and azimuth, and the point of impact by its x and y coordinates in the telescope array reference frame (for the CTA site being concerned, La Palma, for instance).

Moreover, while searching for a model that best captures the physics underlying the reconstruction of gamma events, we also evaluated other, simpler and more complex architectures for this multitasking block. None of them, however, performed better than the one presented here.

7.3.3. *Improving the encoder with attention*

To better adapt our model to the specificity of the analysis of IACT-produced data, we increase the γ-PhysNet encoder with attention. The latter should allow the network to learn how to focus on the relevant features in the data, and thus better take advantage of the signal information buried in the noise of the images. This could also improve how to process truncated showers on the edges of the camera that might disrupt the learning process.

As mentioned earlier, the γ-PhysNet encoder is the convolutional part of the ResNet-56 version Cifar-10. As illustrated in Figure 7.10, it is composed of an initial convolution layer, then three residual stages whose first layer performs subsampling using a convolution.

Since inference time is critical for the analysis of CTA data, it would be costly to insert the attention blocks after each of the 55 convolutions of the encoder. We therefore insert these blocks after each stage in order to make the model take advantage of the attention at each of the size scales of the feature maps, while having a reasonable cost in computing time.

We compared three attention mechanisms for γ-PhysNet in a large study, by varying the hyperparameter controlling their bottleneck: *squeeze-and-excitation*, *self-attention* and *dual attention*. *Squeeze-and-excitation* and *dual attention* have consistently achieved better results than *self-attention*, without being able to tell which is best. In the rest of this chapter, we will focus on *dual attention*, consisting of a channel attention path and a spatial attention path. As part of a research process for model explainability, the observation of the attention maps produced by the latter can indeed give indications on the way how the network makes its decisions. The version of the model augmented with the *dual attention* blocks is called γ-PhysNet DA.

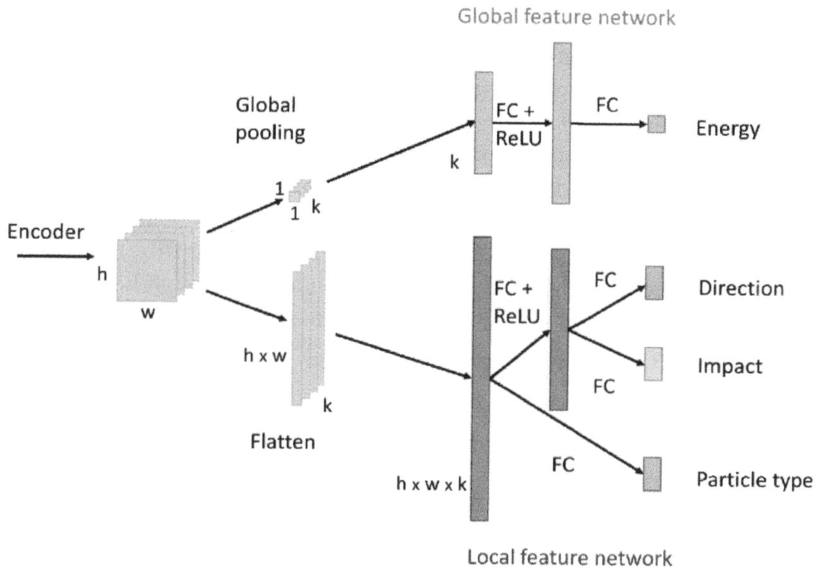

Figure 7.9. *Multitasking block inspired by the physics of reconstruction. FC stands for fully connected layer*

7.4. Performance evaluation

In order to validate our approach, we evaluate the γ-PhysNet performance (without attention) on LST1 simulation data in a simple telescope analysis context. We first demonstrate the contribution of multitasking to reconstruction and compare the results of our architecture with those of the standard Hillas + RF approach. Next, we measure the impact of attention on the behavior of the model, and in particular on its robustness, and we compare two visual explainability methods to understand this impact.

7.4.1. *Dataset*

Since it is impossible to obtain ground truth for actual data in gamma astronomy, the evaluation of the performance of γ-PhysNet is achieved using high-quality simulation thanks to a very good knowledge of the physical model for developing particle showers. Two stages comprise the event simulation. The particle shower is first generated from the parameters of the incident particle using a Monte Carlo method. The tool used is Corsika (Heck et al. 1998). Then, the response of the telescope is simulated using sim_telarray (Bernlöhr 2008) software based on a model of its optical and electronic systems. It is interesting to note that this modeling is updated with the feedback from the telescope operation. The reference dataset used in this chapter is the LST4 mono-trigger Production (as of April 15, 2019). It is the result of large-scale production generated by the LST collaboration for the commissioning of LST1 and contains the data generated for the four LSTs considered for CTA[6]. These data are composed of events of different types, including diffuse gammas, that is, corresponding to extended sources, gammas originating from a point source (named *point-like*) and protons (always diffuse). Gammas and protons are simulated with different energy distributions (although both following a power law of spectral index -2), leading to an unbalanced dataset in terms of the number of events per energy band.

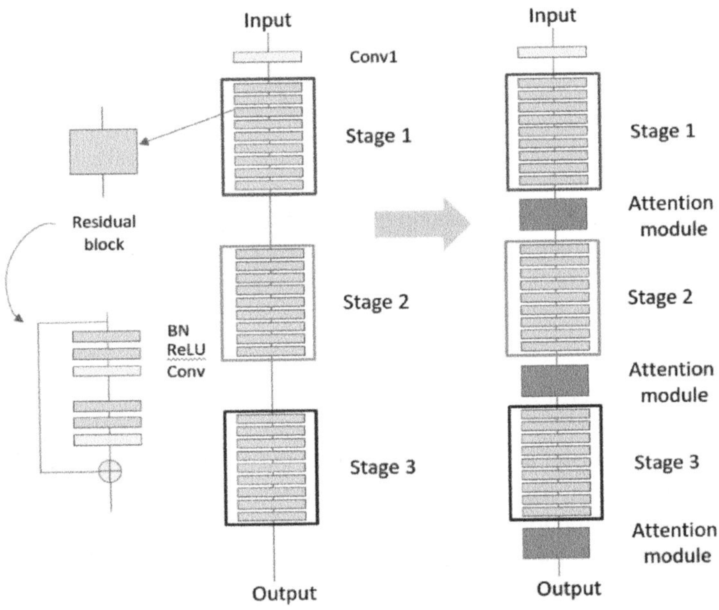

Figure 7.10. *Introduction of attention mechanisms into the γ-PhysNet encoder. The attention modules are inserted after each ResNet-56 stage*

6 Available at: https://www.cta-observatory.org/science/cta-performance/.

7.4.2. *Data selection and preparation*

The first step for the analysis of the standard Hillas + RF method consists of preparing the data to suppress noise in the images and selecting the data to eliminate events at the detectability threshold of LST1, in which the intensity of the Cherenkov light is too low compared to the background noise of the sky and therefore difficult to reconstruct. Conventionally, this selection is based on three criteria: the size of the signal (in number of pixels) after noise suppression, its total intensity and the fraction of the signal located on the edge of the camera, indicating whether it is truncated or not.

We follow the same approach in order to be able to compare our models to the Hillas + RF reference, except that we do not remove noise from images. The strong point of deep learning is indeed to extract fine information from all the pixels of the image. As we have experienced, removing noise from images limits the model performance to the quality of the deletion.

We then consider common training, validation and testing databases.

7.4.3. *Model training*

For the experiments presented later in this chapter, all models (neural networks and Hillas + RF) are trained using the four telescopes in the dataset to get a more accurate overview of data variability. The models are trained on diffuse gammas to be capable of reconstructing events arriving from all directions in the field of view of telescopes and on proton events.

All neural networks are trained using the same hyperparameters. Indeed, because of the duration of the training (between 10 and 40 h, depending on the selections applied to the data), it is not feasible at this stage of the study to perform an advanced optimization of each model. Nevertheless, starting from the hyperparameters optimized for the ResNet-56, substantial preliminary experiments have led to identifying a set of high-performance common hyperparameters utilized for the experiments presented below and allowing each model to achieve its best level of performance.

Specifically, neural networks are trained on 25 epochs using the Adam optimizer (Kingma and Ba 2015). The learning rate is set at 10^{-3} and is divided by 10 every 10 epochs. The learning is regularized by applying a penalty of 10^{-4} on the L2 norm of the weights in the model.

As pointed out in section 7.2.2, it is necessary for multitasking networks to balance tasks. We use the method proposed by Kendall et al. (2018) and based on the estimation of the uncertainty related to each task. It actually gives the best results for our problem. Since task uncertainty is a learned parameter, we use the Adam optimizer again, this time with a learning rate of 0.025 and a penalty on the weight norm of 10^{-4}.

For the classification task (gamma/proton separation), we use classical cross-entropy as the optimization criterion. For regression tasks (energy, direction and virtual point of impact), the selected criterion is an L1 distance (absolute mean error) which yielded better performance than the mean squared error (classical L2 distance) in our preliminary experiments. In addition, in order to prevent protons from disturbing the learning of gamma energy, direction and point of impact, we use a strategy known as masking (*masked loss*). This consists of zeroing the error of regression tasks computed by the cost function when the particle is a proton:

$$L_t = \frac{1}{M} \sum_N \begin{cases} l_i, & \text{if the particle is a gamma} \\ 0, & \text{otherwise} \end{cases} \qquad [7.1]$$

with L_t the task error t for a data batch of size N, a number of gamma events M in the batch, and l_i the error for the example i.

7.4.4. *Performance evaluation methodology*

To comply to standard practice in gamma astronomy, the different models are evaluated on gammas originating from a single-point source (*point-like*) and protons; this configuration corresponds to the conventional use cases of the field.

The performance of the different models for regression tasks (energy and direction) is measured using resolution curves. For energy, the resolution represents, per energy band, the width of the interval around 0 containing 68% of the distribution of the relative error made by the model during prediction. On the other hand, the angular resolution represents, by energy band, the angle in which 68% of the reconstructed gammas fall, relative to their true direction. For both resolution curves, lower values indicate better results.

For the gamma/proton classification task, the overall performance of the models is expressed by the area under the ROC curve (AUC), while the precision indicates the ability of the models to reject protons, and the recall illustrates their ability to retain gammas. Precision and recall are calculated for a default threshold value of 0.5 for the gamma output of the model.

Within the framework of deep learning, stochastic initialization of neural network weights plays an important role in the final performance of the trained model. In order to have a performance estimate that is less biased and not dependent on a particular random seed, we repeat the training of all neural networks six times with different seeds. This also makes it possible to measure the variability of the models due to the initialization of the parameters and therefore their robustness. To account for this variability, resolution curves are plotted as bands representing two standard deviations around the mean of the results illustrated by points. This average resolution has no physical reality, since resolution is a statistical measure of the error of a particular model. However, it provides a trend of model performance and is useful for the readability of the results.

7.4.5. *Interest of multitasking and comparison with the standard method*

To assess the interest of multitasking deep learning for Cherenkov data analysis, namely gamma/proton classification and energy and direction regression, we compare the performance of the following approaches.

– γ-PhysNet: the architecture that we propose.

– γ-PhysNet w/o impact: to evaluate the contribution of the point of impact as an auxiliary task, we train γ-PhysNet after removing this task.

– ResNet-56: to demonstrate the contribution of multitasking learning to the performance of γ-PhysNet, we train three ResNet-56s, corresponding to the γ-PhysNet encoder, to solve the different tasks. The main difference with the architecture described by He et al. (2016b) can be seen in the fact that for direction regression and gamma/proton classification tasks, we replace the final global mean preceding the last fully connected layer with a vectorization of the features produced by the encoder in order to preserve spatial information.

– Hillas + RF: this widely used method for reconstructing gamma events from Cherenkov data is used as a reference.

We cannot compare our approach with those proposed by Holch et al. (2017), Shilon et al. (2019), Parsons and Ohm (2020) and Mangano et al. (2018), because these are designed for stereoscopic analysis of Cherenkov data, when our architecture performs analysis from data from a single telescope. We also cannot compare our results with those of Nieto et al. (2017), because the latter were obtained for a different type of telescope, only for the classification task and without using the temporal information of the data.

For the study presented in this chapter, we apply a selection of data as described in section 7.4.2. More specifically, events with the following characteristics are retained:

– total intensity greater than 300 photoelectrons;

– at least one pixel survives the noise suppression operation;

– the signal fraction, in intensity, located on the two rows bordering the image, is less than 20%.

These cuts are standard in gamma-ray astronomy, although a bit strong at low energy. The result is a training game consisting of $388k$ diffuse gammas and $236k$ protons.

7.4.5.1. *Gamma/proton classification*

The results reported in Table 7.1 clearly show that, with the same order of precision, all neural networks yield much better recalls than the standard Hillas + RF method. In particular, the proposed architecture improves the recall by 60.1%. This improvement will make it possible to augment the sensitivity of the analysis by retaining in addition 60% of gamma events. The contribution of multitasking is also significant compared to the single-task approach achieved with the ResNet-56 model, with once again an improvement in recall. However, introducing the impact point regression as an auxiliary task does not seem to benefit the gamma/proton classification task.

Model	AUC	Precision	Recall
Hillas + RF	0.898	0.956	0.593
ResNet-56	0.954±0.001	0.956±0.001	0.942±0.001
γ-PhysNet	*0.960±0.002*	0.957±0.003	*0.956±0.006*
γ-PhysNet w/o impact	*0.961±0.002*	*0.960±0.001*	0.949±0.003

Table 7.1. *AUC score, precision and gamma/proton classification task recall for different models*

7.4.6. *Energy regression*

The results presented in Figure 7.11 show that, as with the classification task, all neural network-based models perform better than the standard Hillas + RF method. Furthermore, γ-PhysNet particularly reduces the energy error to 35% at high energy, and up to 75% at 31 GeV (the point of the Hillas + RF curve is outside the figure). At energies above 400 GeV, the no-impact multitasking approach degrades performance in terms of resolution and dispersion. This can be explained by the fact that the light intensity emitted by the electromagnetic shower is strongly correlated with the energy of the detected particle. The intensity detected by the telescope then depends on its distance from the virtual point of impact of the particle. The addition of impact point regression as an auxiliary task enables γ-PhysNet to achieve a performance comparable to the single-task approach performed with the ResNet-56; however, with greater scattering at high energies.

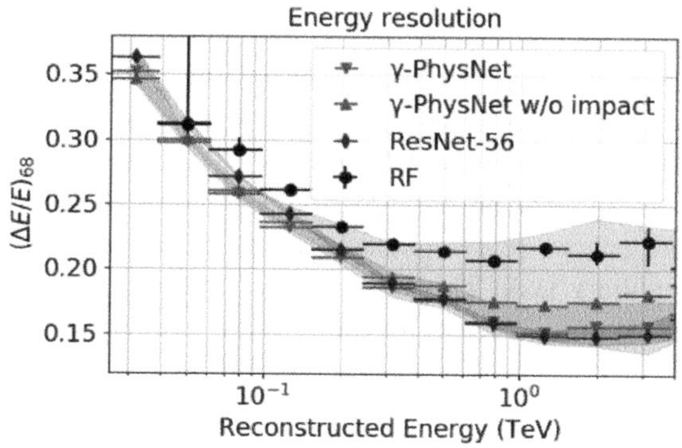

Figure 7.11. *Energy resolution as a function of energy rebuilt in the LST1 energy band (lower values indicate a better result). Performance comparison of the various different models*

7.4.7. Direction regression

As with the gamma/proton classification and energy regression tasks, Figure 7.12 shows that neural networks perform better with the direction regression task than the Hillas + RF method. More specifically, γ-PhysNet improves the reconstruction up to values ranging from 0.03 to 0.3° depending on the energy being considered. In addition, this time, the multitasking approach reduces the reconstruction error by about 0.08° at low energy. Both multitasking approaches achieve similar results, with the addition of impact point regression for reducing the variability created by initializing the weights.

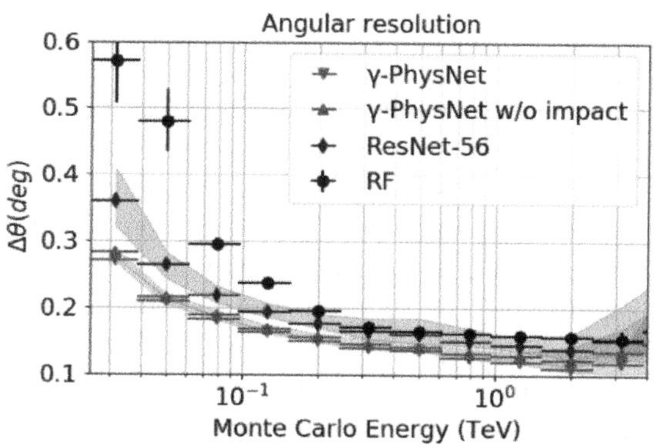

Figure 7.12. *Angular resolution according to the simulated energy in the LST1 energy band (lower values indicate a better result). Performance comparison with different models*

7.4.7.1. *Conclusion*

The analysis of the results for the three tasks of gamma event reconstruction from Cherenkov data shows that the proposed architecture clearly achieves better performance than the standard Hillas + RF method. The multitasking approach makes it possible to reduce the error on the direction regression while maintaining similar performance on the energy regression as a result of introducing the virtual point of impact as an auxiliary task. This addition also leads to reducing the variability introduced by initializing parameters for multitasking models.

7.4.8. *Impact of attention*

As stated in section 7.3, we propose to increase γ-PhysNet with *dual attention* blocks. The objective is to introduce devices for the regularization and adaptation of the model to different contexts of analysis, and also to introduce prediction explainability through the analysis of the attention maps produced for each prediction. In this section, we evaluate the effective contribution of these mechanisms by comparing the performance of γ-PhysNet and γ-PhysNet DA for the three tasks of gamma event reconstruction from Cherenkov data: gamma/proton classification, energy and direction regression. Since this time, we are not comparing ourselves to the standard Hillas + RF method, we can relax the data selection in order to extend the energy range of interest with respect to the experiments presented in section 7.4.5. In other words, events having a total intensity greater than 50 photoelectrons are preserved, compared to 300 previously. By retaining more low-energy events conventionally rejected by standard analysis, we pave the way to improving the sensitivity of the analysis to address relevant use cases such as the study of extragalactic objects and transient sources. With this selection, the training dataset now comprises $874k$ gammas and $506k$ protons.

Average performance metrics on the classification task are indicated in Table 7.2. We find that the use of attention does not lead to improving the results of the classification task: performance is identical, with or without attention. Nonetheless, for the direction and energy regression tasks, we can see in Figure 7.13 that γ-PhysNet DA achieves significantly better results at high energy. Mainly, result heterogeneity can significantly be reduced using attention. In particular, for the energy regression task, scattering with attention is up to three times less above 200 GeV. Moreover, this lower result scattering from γ-PhysNet DA on the two regression tasks indicates better robustness to initializing model parameters. This will yield a more reliable estimate of the parameters of the particle detected during the analysis of the actual data produced by LST1.

Model	AUC	Precision	Recall
γ-PhysNet	0.882±0.001	0.929±0.001	0.935±0.007
γ-PhysNet DA	0.882±0.001	0.929±0.001	0.935±0.005

Table 7.2. *AUC score, precision and recall of the gamma/proton classification task for γ-PhysNet with (DA) and without attention*

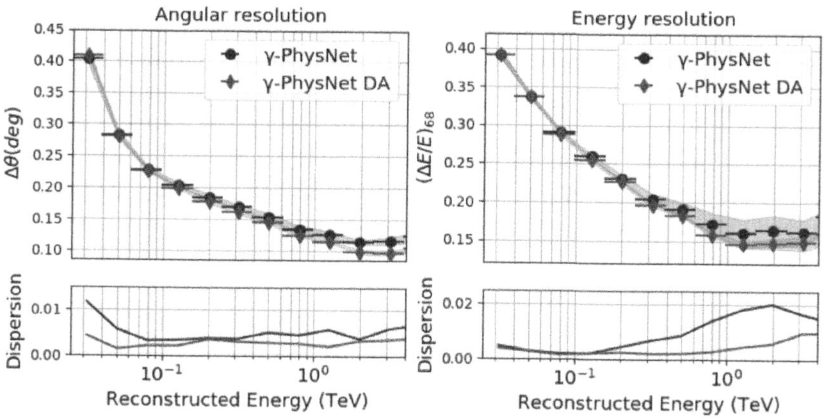

Figure 7.13. *(a) Angular and (b) energy resolutions according to the energy rebuilt in the LST1 energy band (lower values indicate a better result)*

COMMENT ON FIGURE 7.13.– *These curves represent the model error, respectively, for the direction and energy regression of the detected gamma ray. Scattering, representing the variability induced by initializing weights, is a measure of the robustness of the model.*

7.4.9. *Understanding the effect of attention on robustness*

In order to understand why adding *dual attention* blocks to γ-PhysNet has no impact on the classification task, but improves the robustness of energy and direction regression tasks, we implement methods to explain the predictions described in section 7.2.4. We first carefully analyze the Grad-CAM heat maps produced by γ-PhysNet and γ-PhysNet DA for 25 well and badly reconstructed events from the test set. We also observe the spatial attention maps of the three blocks *dual attention*. In addition, we produce a global spatial attention map of the model by combining the attention maps using the Hadamard product. Finally, since the models are trained with six random seeds, we calculate the average of the Grad-CAM heat maps and spatial attention maps for each pixel, for a particular event.

7.4.9.1. *Analysis of Grad-CAM heat maps*

Figure 7.14 depicts a representative event of the general trend observed in the 25 events studied. The first learning concerns the task of classification. The observation of heat maps with and without attention shows that, in both cases, the most important pixels for model decision are located in the signal area. However, γ-PhysNet DA (with attention) takes a wider region around it into account.

For energy regression, we observe that without attention, the network mainly favors the pixels located on the edges of the camera. From a physical point of view, this can make sense, because the fact that the signal is truncated or not is an important insight for the reconstruction of energy. For a given direction and point of impact, the amount of signal received by the telescope camera is indeed approximately proportional to the energy of the gamma ray detected. With γ-PhysNet DA (with attention), we notice that the model focuses this time on a large area around the signal.

Analysis of altitude and azimuth heat maps shows similar behavior for γ-PhysNet without attention. Once again, whether a signal is truncated or not impacts the reconstruction of the direction, in particular by altering the information on the development of the shower given by the channel of the images containing the temporal information. In turn, the model with attention focuses on the signal area, more particularly on the ends of the shower.

7.4.9.2. *Observing spatial attention maps*

By observing the spatial attention maps of γ-PhysNet DA, it can be seen that the effect of the *dual attention* blocks is different depending on the scale of the feature maps. At the output of stage 1 of the encoder, the feature maps are of the same resolution as the input data of the model. At this scale, the attention mechanism strongly highlights the signal pixels. As the *dual attention* modules produce weights in the $[1; 2]$ range, the signal area at this scale is given a weight of 1.8. At this point, however, the attention maps are quite noisy. After stage 2 of the encoder, spatial attention maps also strongly highlight the signal area in the data. However, they are less noisy. We then observe that the last attention module, after stage 3, has a lower impact on the values of the feature maps. Finally, the analysis of the combination of the 3 attention maps brings forward that attention strongly aids the model to focus on the signal pixels, which is consistent with the observation of the Grad-CAM heat maps.

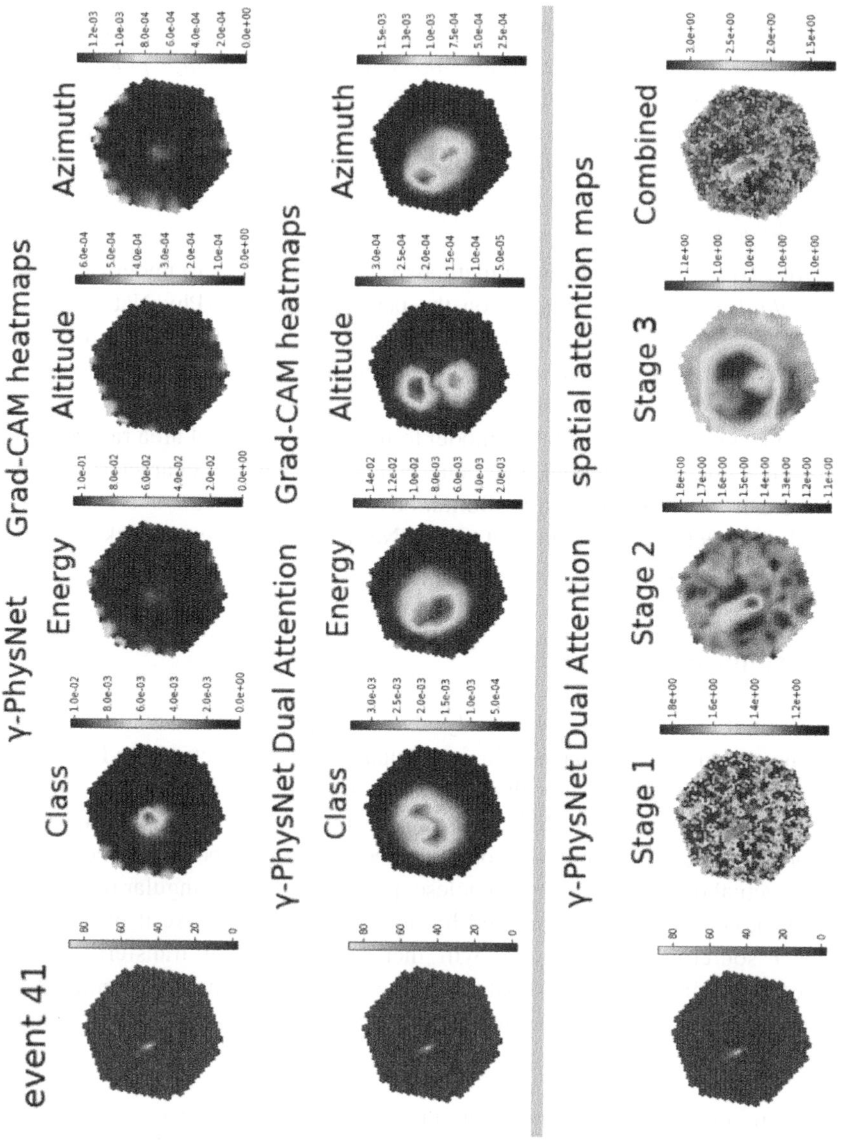

Figure 7.14. *Behavior analysis of* γ*-PhysNet with and without attention for a representative event*

COMMENT ON FIGURE 7.14.– *The first column shows the analyzed event (only the channel containing the intensity of each pixel). The first row contains the Grad-CAM heat maps produced with the model without attention for each of the tasks (gamma/proton classification, energy regression and direction regression in terms of altitude and azimuth). The second row shows the heat maps for the model with attention. The third row carefully represents the spatial attention maps placed after each of the three stages of the model, as well as the combination of these three maps. Since each model is trained with six different random seeds, the different maps are averaged per pixel over the six versions of each model.*

7.4.9.3. *Discussion*

The use of two visual explainability methods makes it possible to understand the effect of the *dual attention* mechanism on the performance of γ-PhysNet. For the gamma/proton classification task, both model versions with and without attention focus on the area of the image containing the signal, which may explain their identical results with this task, both in terms of performance and variability. In contrast, for the two regression tasks, attention leads the model to focus on the signal area rather than the edges of the camera. The model then better analyzes the actual characteristics of the signal and can overcome the spatial limitations of the sensor. This is a clue as to explain the slightly better performance of γ-PhysNet DA with these two tasks and its greater robustness to initialize network weights.

7.5. Conclusions and outlook

The deep multitasking approach clearly shows its potential for reconstructing gamma events from IACT images. In particular, our architecture, γ-PhysNet, achieves significantly better results than the standard Hillas + RF method for the analysis of simulation data from LST1 used to prepare analysis tools. The improved energy resolution should allow for more detailed spectra to be produced when analyzing the actual data produced by the telescope. The improved angular resolution and gamma/proton classification provided by our approach will make it possible to detect weaker sources. The next step will therefore consist of transferring this performance to the analysis of actual data, because despite the simulations employed for model training being of high quality, the data produced by LST1 will certainly differ.

With this in mind, adding blocks of attention to our architecture results in improving its robustness by reducing the variability induced by stochastic initialization of network parameters. In effect, the use and comparison of two visual explainability methods show that the *dual attention* mechanism helps the model to better focus on the area of the image containing the signal. In order to take the interpretability of the model to a new level, it will probably be necessary to build, from these visual methods, statistical indicators for providing a systematic performance analysis, both on simulation data and real data.

7.6. Acknowledgments

We gratefully acknowledge the financial support of the agencies and organizations listed here: www.cta-observatory.org/consortium_acknowledgment. This project was funded by the *European Union's Horizon 2020 research and innovation program* on the basis of the grant agreement No. 653477, and the Foundation of the Université Savoie Mont Blanc. This research was carried out by virtue of the services offered by the Must meso-center of the Université Savoie Mont Blanc – CNRS/IN2P3, of the computing resources of Idris through the resource allocation 2020-AD011011577 allocated by GENCI, of the computing and data processing resources of the CNRS/IN2P3 Computing Center (Lyon, France). We gratefully acknowledge Nvidia Corporation's support for donating an Nvidia P6000 GPU for this research.

7.7. References

Albert, J., Aliu, E., Anderhub, H., Antoranz, P., Armada, A., Asensio, M., Baixeras, C., Barrio, J., Bartko, H., Bastieri, D. et al. (2008). Implementation of the random forest method for the imaging atmospheric Cherenkov telescope magic. *Nuclear Instruments and Methods in Physics Research Section A; Accelerators, Spectrometers, Detectors and Associated Equipment*, 588(3), 424–432.

Bahdanau, D., Cho, K., Bengio, Y. (2015). Neural machine translation by jointly learning to align and translate. In *3rd International Conference on Learning Representations, ICLR 2015*, May 7–9, San Diego.

Baxter, J. (1997). A Bayesian/information theoretic model of learning to learn via multiple task sampling. *Machine Learning*, 28(1), 7–39.

Bernlöhr, K. (2008). Simulation of imaging atmospheric Cherenkov telescopes with CORSIKA and sim_telarray. *Astroparticle Physics*, 30(3), 149–158.

Bock, R., Chilingarian, A., Gaug, M., Hakl, F., Hengstebeck, T., Jiřina, M., Klaschka, J., Kotrč, E., Savickỳ, P., Towers, S. et al. (2004). Methods for multidimensional event classification; a case study using images from a cherenkov gamma-ray telescope. *Nuclear Instruments and Methods in Physics Research Section A; Accelerators, Spectrometers, Detectors and Associated Equipment*, 516(2/3), 511–528.

Boynton, G.M. (2005). Attention and visual perception. *Current Opinion in Neurobiology*, 15(4), 465–469.

Brunel, A., Pasquet, J., Rodriguez, N., Comby, F., Fouchez, D., Chaumont, M. (2019). A CNN adapted to time series for the classification of supernovae. *Electronic Imaging*, 14, 90–91.

Cao, C., Liu, X., Yang, Y., Yu, Y., Wang, J., Wang, Z., Huang, Y., Wang, L., Huang, C., Xu, W. et al. (2015). Look and think twice; Capturing top-down visual attention with feedback convolutional neural networks. In *Proceedings of the IEEE International Conference on Computer Vision*, Santiago.

Cao, J., Li, Y., Zhang, Z. (2018). Partially shared multi-task convolutional neural network with local constraint for face attribute learning. In *Proceedings of the IEEE Conference on Computer Vision and Pattern Recognition*, Salt Lake City.

Caruana, R. (1997). Multitask learning. *Machine Learning*, 28(1), 41–75.

Chen, L., Zhang, H., Xiao, J., Nie, L., Shao, J., Liu, W., Chua, T.-S. (2017). SCA-CNN; Spatial and channel-wise attention in convolutional networks for image captioning. In *Proceedings of the IEEE Conference on Computer Vision and Pattern Recognition*, Stockholm.

Chen, Z., Badrinarayanan, V., Lee, C.-Y., Rabinovich, A. (2018). GradNorm: Gradient normalization for adaptive loss balancing in deep multitask networks. In *Proceedings of the 35th International Conference on Machine Learning*. Association for Computational Linguistics, Austin.

Cheng, J., Dong, L., Lapata, M. (2016). Long short-term memory-networks for machine reading. In *Proceedings of the 2016 Conference on Empirical Methods in Natural Language Processing*. Association for Computational Linguistics, Minneapolis.

Devlin, J., Chang, M.-W., Lee, K., Toutanova, K. (2019). BERT; Pre-training of deep bidirectional transformers for language understanding. In *Proceedings of the 2019 Conference of the North American Chapter of the Association for Computational Linguistics; Human Language Technologies*.

Feng, Q., Lin, T.T., for the Veritas Collaboration (2016). The analysis of VERITAS muon images using convolutional neural networks. *Proceedings of the International Astronomical Union*, 12(S325), 173–179.

Fiasson, A., Dubois, F., Lamanna, G., Masbou, J., Rosier-Lees, S. (2010). Optimization of multivariate analysis for IACT stereoscopic systems. *Astroparticle Physics*, 34.

Guo, M., Haque, A., Huang, D.-A., Yeung, S., Fei-Fei, L. (2018). Dynamic task prioritization for multitask learning. In *Proceedings of the European Conference on Computer Vision (ECCV)*. Springer, Munich.

Han, H., Jain, A.K., Wang, F., Shan, S., Chen, X. (2017). Heterogeneous face attribute estimation; A deep multi-task learning approach. *IEEE Transactions on Pattern Analysis and Machine Intelligence*, 40(11), 2597–2609.

He, K., Zhang, J., Ren, S., Sun, J. (2016a). Identity mappings in deep residual networks. In *European cConference on Computer Vision*. Springer, Amsterdam.

He, K., Zhang, X., Ren, S., Sun, J. (2016b). Deep residual learning for image recognition. In *Proceedings of the IEEE Conference on Computer Vision and Pattern Recognition*, Las Vegas.

Heck, D., Knapp, J., Capdevielle, J., Schatz, G., Thouw, T. (1998). CORSIKA: A Monte Carlo code to simulate extensive air showers. *Report FZKA*, 6019(11).

Hillas, A. (1985). Cerenkov light images of eas produced by primary gamma. *International Cosmic Ray Conference*, San Diego.

Holch, T.L., Shilon, I., Büchele, M., Fischer, T., Funk, S., Groeger, N., Jankowsky, D., Lohse, T., Schwanke, U., Wagner, P. (2017). Probing convolutional neural networks for event reconstruction in gamma-ray astronomy with Cherenkov telescopes. In *35th International Cosmic Ray Conference (ICRC2017)*, Busan.

Hu, J., Shen, L., Sun, G. (2018). Squeeze-and-excitation networks. In *Proceedings of the IEEE Conference on Computer Vision and Pattern Recognition*, Salt Lake City.

Huang, G., Liu, Z., van der Maaten, L., Weinberger, K.Q. (2017). Densely connected convolutional networks. *2017 IEEE Conference on Computer Vision and Pattern Recognition (CVPR)*, Honolulu.

Iizuka, S., Simo-Serra, E., Ishikawa, H. (2016). Let there be color! Joint end-to-end learning of global and local image priors for automatic image colorization with simultaneous classification. *ACM Transactions on Graphics (ToG)*, 35(4), 1–11.

Jacquemont, M., Antiga, L., Vuillaume, T., Silvestri, G., Benoit, A., Lambert, P., Maurin, G. (2019). Indexed operations for non-rectangular lattices applied to convolutional neural networks. *Proceedings of the 14th International Joint Conference on Computer Vision, Imaging and Computer Graphics Theory and Applications*, 5, 362–371.

Kendall, A., Gal, Y., Cipolla, R. (2018). Multi-task learning using uncertainty to weigh losses for scene geometry and semantics. In *Proceedings of the IEEE Conference on Computer Vision and Pattern Recognition*, Salt Lake City.

Kim, E.J. and Brunner, R.J. (2016). Star-galaxy classification using deep convolutional neural networks. *Monthly Notices of the Royal Astronomical Society*, 464(4), stw2672.

Kim, B., Humensky, B., Nieto, D., Brill, A., Desai, M., for the Cherenkov Telescope Array Collaboration (2017). Exploring deep learning as an event classification method for the Cherenkov telescope array. *arXiv*, 1–8.

Kingma, D.P. and Ba, J. (2015). Adam; A method for stochastic optimization. *3rd International Conference on Learning Representations, ICLR 2015*, May 7–9, San Diego.

Kokkinos, I. (2017). UberNet: Training a universal convolutional neural network for low-, mid-, and high-level vision using diverse datasets and limited memory. In *Proceedings of the IEEE Conference on Computer Vision and Pattern Recognition*, Honolulu.

Krizhevsky, A., Sutskever, I., Hinton, G.E. (2012). Imagenet classification with deep convolutional neural networks. In *Advances in Neural Information Processing Systems*. Curran Associates Inc., Lake Tahoe.

López-Coto, R., Baquero, A., Bernardos, M.I., Cassol, F., Foffano, L., García, E., Gliwny, P., Iwamura, Y., Jacquemont, M., Jouvin, L. et al. (2021). lstchain: An analysis pipeline for LST-1, the first prototype large-sized telescope of CTA. In *Astronomical Data Analysis Software and Systems XXX, Astronomical Society of the Pacific Conference Series*, Astronomical Society of the Pacific.

Luvizon, D.C., Picard, D., Tabia, H. (2018). 2D/3D pose estimation and action recognition using multitask deep learning. In *Proceedings of the IEEE Conference on Computer Vision and Pattern Recognition*, Salt Lake City.

Mangano, S., Delgado, C., Bernardos, M.I. Lallena, M., Rodriguez Vazquez, J.J., for the CTA Consortium (2018). Extracting gamma-ray information from images with convolutional neural network methods on simulated Cherenkov telescope array data. In *IAPR Workshop on Artificial Neural Networks in Pattern Recognition*. Springer, Amsterdam.

Morcos, A.S., Barrett, D.G.T., Rabinowitz, N.C., Botvinick, M. (2018). On the importance of single directions for generalization. *6th International Conference on Learning Representations, ICLR 2018*, April 30–May 3, Vancouver.

de Naurois, M. and Rolland, L. (2009). A high performance likelihood reconstruction of γ-rays for imaging atmospheric Cherenkov telescopes. *Astroparticle Physics*, 32(5), 231–252.

Nieto, D., Brill, A., Feng, Q., Jacquemont, M., Kim, B., Miener, T., Vuillaume, T. (2019). Studying deep convolutional neural networks with hexagonal lattices for imaging atmospheric Cherenkov telescope event reconstruction. *ICRC 2019 - 36th International Cosmic Ray Conference*, Madison.

Ohm, S., van Eldik, C., Egberts, K. (2009). γ/hadron separation in very-high-energy γ-ray astronomy using a multivariate analysis method. *Astroparticle Physics*, 31(5), 383–391.

Olah, C., Mordvintsev, A., Schubert, L. (2017). Feature visualization. *Distill*, 2(11), e7.

Parmar, N., Vaswani, A., Uszkoreit, J., Kaiser, L., Shazeer, N., Ku, A., Tran, D. (2018). Image transformer. In *Proceedings of the 35th International Conference on Machine Learning*, Stockholm.

Parmar, N., Ramachandran, P., Vaswani, A., Bello, I., Levskaya, A., Shlens, J. (2019). Stand-alone self-attention in vision models. In *Advances in Neural Information Processing Systems 32*, Wallach, H., Larochelle, H., Beygelzimer, A., d'Alché-Buc, F., Fox, E., Garnett, R. (eds). Curran Associates, Inc., Red Hook.

Parsons, R.D. and Hinton, J.A. (2014). A Monte Carlo template based analysis for air-Cherenkov arrays. *Astroparticle Physics*, 56, 26–34.

Parsons, R.D. and Ohm, S. (2020). Background rejection in atmospheric Cherenkov telescopes using recurrent convolutional neural networks. *European Physical Journal C*, 80(5), 363.

Ren, Z. and Jae Lee, Y. (2018). Cross-domain self-supervised multi-task feature learning using synthetic imagery. In *Proceedings of the IEEE Conference on Computer Vision and Pattern Recognition*, Salt Lake City.

Russakovsky, O., Deng, J., Su, H., Krause, J., Satheesh, S., Ma, S., Huang, Z., Karpathy, A., Khosla, A., Bernstein, M. et al. (2015). ImageNet large scale visual recognition challenge. *International Journal of Computer Vision (IJCV)*, 115(3), 211–252.

Saenz, M., Buracas, G.T., Boynton, G.M. (2002). Global effects of feature-based attention in human visual cortex. *Nature Neuroscience*, 5(7), 631–632.

Scientific Figure on ResearchGate (2009). Analysis methods for Gamma-ray astronomy [Onine]. Available at: https://www.researchgate.net/figure/llustration-of-the-stereoscopic-event-reconstruction-method-The-shower-direction-is_fig5_373715823 [Accessed 8 March, 2024].

Selvaraju, R.R., Cogswell, M., Das, A., Vedantam, R., Parikh, D., Batra, D. (2017). Grad-CAM: Visual explanations from deep networks via gradient-based localization. In *Proceedings of the IEEE International Conference on Computer Vision*, Venice.

Sener, O. and Koltun, V. (2018). Multi-task learning as multi-objective optimization. In *Advances in Neural Information Processing Systems*, Montréal.

Shilon, I., Kraus, M., Büchele, M., Egberts, K., Fischer, T., Holch, T.L., Lohse, T., Schwanke, U., Steppa, C., Funk, S. (2019). Application of deep learning methods to analysis of imaging atmospheric Cherenkov telescopes data. *Astroparticle Physics*, 105, 44–53.

Simonyan, K. and Zisserman, A. (2015). Very deep convolutional networks for large-scale image recognition. In *3rd International Conference on Learning Representations, ICLR 2015*, San Diego.

Springenberg, J.T., Dosovitskiy, A., Brox, T., Riedmiller, M.A. (2015). Striving for simplicity; The all convolutional net. In *3rd International Conference on Learning Representations, ICLR 2015*, San Diego.

Srinivas, S. and Fleuret, F. (2019). Full-gradient representation for neural network visualization. In *Advances in Neural Information Processing Systems*. Neural Information Processing Systems Foundation Inc., Vancouver.

Sun, J., Darbeha, F., Zaidi, M., Wang, B. (2020). SAUNet: Shape attentive U-net for interpretable medical image segmentation. *arXiv*.

Szegedy, C., Liu, W., Jia, Y., Sermanet, P., Reed, S., Anguelov, D., Erhan, D., Vanhoucke, V., Rabinovich, A., Hill, C. et al. (2015). Going deeper with convolutions. In *Proceedings of the IEEE Conference on Computer Vision and Pattern Recognition*, Boston.

Touvron, H., Vedaldi, A., Douze, M., Jégou, H. (2019). Fixing the train-test resolution discrepancy. In *Advances in Neural Information Processing Systems*. Neural Information Processing Systems Foundation Inc., Vancouver.

Vaswani, A., Shazeer, N., Parmar, N., Uszkoreit, J., Jones, L., Gomez, A.N., Kaiser, Ł., Polosukhin, I. (2017). Attention is all you need. In *Advances in Neural Information Processing Systems*. Neural Information Processing Systems Foundation Inc., Long Beach.

Völk, H.J., Bernlöhr, K. (2009). Imaging very high energy gamma-ray telescopes. *Experimental Astronomy*, 25(1–3), 173–191.

Wang, X., Girshick, R., Gupta, A., He, K. (2018). Non-local neural networks. In *Proceedings of the IEEE Conference on Computer Vision and Pattern Recognition*, Salt Lake City.

Weekes, T.C., Cawley, M.F., Fegan, D., Gibbs, K., Hillas, A., Kowk, P., Lamb, R., Lewis, D., Macomb, D., Porter, N. et al. (1989). Observation of tev gamma-rays from the crab nebula using the atmospheric Cherenkov imaging technique. *Astrophysical Journal*, 342, 379–395.

Woo, S., Park, J., Lee, J.-Y., So Kweon, I. (2018). CBAM: Convolutional block attention module. In *Proceedings of the European Conference on Computer Vision (ECCV)*. Springer, Munich.

Xu, D., Ouyang, W., Wang, X., Sebe, N. (2018). PAD-Net; Multi-tasks guided prediction-and-distillation network for simultaneous depth estimation and scene parsing. In *Proceedings of the IEEE Conference on Computer Vision and Pattern Recognition*, Salt Lake City.

Yang, Y. and Hospedales, T.M. (2017). Trace norm regularised deep multi-task learning. *5th International Conference on Learning Representations, ICLR 2017*, April 24–26, Toulon.

Yuan, Y., Chen, X., Wang, J. (2020). Object-contextual representations for semantic segmentation. In *Computer Vision – ECCV 2020*. Springer, Glasgow.

Zhang, H., Goodfellow, I., Metaxas, D., Odena, A. (2019). Self-attention generative adversarial networks. In *International Conference on Machine Learning*, Long Beach.

Zhang, H., Wu, C., Zhang, Z., Zhu, Y., Lin, H., Zhang, Z., Sun, Y., He, T., Mueller, J., Manmatha, R. et al. (2022). Resnest: Split-attention networks. In *Proceedings of the IEEE/CVF conference on computer vision and pattern recognition*, New Orleans, 2736–2746.

Zhou, B., Sun, Y., Bau, D., Torralba, A. (2018). Revisiting the importance of individual units in CNNs via ablation. *arXiv*.

List of Authors

Touria BAJJOUK
Ifremer
DYNECO
LEBCO
Plouzané
France

Mary Grace BATO
JPL
Caltech
Pasadena
USA

Alexandre BENOIT
LISTIC
Université Savoie Mont Blanc
Annecy
France

Malik CHAMI
LATMOS, CNRS
Sorbonne Université
Paris
France

Elise COLIN
ONERA
Université Paris-Saclay
Palaiseau
France

Alexandrine GESRET
Centre de Géosciences
Mines Paris – PSL
Fontainebleau
France

Fabien GILLET-CHAULET
IGE
Université Grenoble Alpes
France

Mikaël JACQUEMONT
LAPP
Université Grenoble Alpes
Annecy
and
LISTIC
Université Savoie Mont Blanc
Annecy
France

Patrick LAMBERT
LISTIC
Université Savoie Mont Blanc
Annecy
France

Gilles MAURIN
LAPP
Université Grenoble Alpes
Université Savoie Mont Blanc
Annecy
France

Audrey MINGHELLI
LIS, CNRS
Université de Toulon
France

Tristan PETIT
Norwegian Polar Institute
Tromsø
Norway

Virginie PINEL
ISTerre
IRD, CNRS
Université Grenoble Alpes
Université Savoie Mont Blanc
Le Bourget-du-Lac
France

Olivier TALAGRAND
LMD/IPSL
CNRS, ENS
Université PSL
Institut polytechnique de Paris
Sorbonne Université
Paris
France

Laetitia THIRION-LEFEVRE
SONDRA
Université Paris-Saclay
Palaiseau
France

Thomas VUILLAUME
LAPP
Université Grenoble Alpes
Université Savoie Mont Blanc
Annecy
France

Yajing YAN
LISTIC
Université Savoie Mont Blanc
Annecy
France

Index

A, B

C, D

E, F

G, H

I, J

L, M

N, O

P, R

S, T

U, V

Printed and bound by CPI Group (UK) Ltd, Croydon, CR0 4YY

13/09/2026

14961068-0001